Collecting and Sorting Data

Survey Questions and Secret Rules

Grade 1

Tracey Wright
Jan Mokros

Contributing Authors
Susan Jo Russell
Antonia Stone

Developed at TERC, Cambridge, Massachusetts

Dale Seymour Publications®

Some material in this unit was developed by Susan Jo Russell and Antonia Stone for *Counting: Ourselves and Our Families* (a unit in the series *Used Numbers: Real Data in the Classroom),* © 1990 by Dale Seymour Publications®.

The *Investigations* curriculum was developed at TERC (formerly Technical Education Research Centers) in collaboration with Kent State University and the State University of New York at Buffalo. The work was supported in part by National Science Foundation Grant No. ESI-9050210. TERC is a nonprofit company working to improve mathematics and science education. TERC is located at 2067 Massachusetts Avenue, Cambridge, MA 02140.

This project was supported, in part, by the
National Science Foundation
Opinions expressed are those of the authors and not necessarily those of the Foundation

This book is published by Dale Seymour Publications®, an imprint of Addison Wesley Longman, Inc.

Managing Editor: Catherine Anderson
Series Editor: Beverly Cory
ESL Consultant: Nancy Sokol Green
Production/Manufacturing Director: Janet Yearian
Production/Manufacturing Manager: Karen Edmonds
Production/Manufacturing Coordinator: Shannon Miller
Design Manager: Jeff Kelly
Design: Don Taka
Composition: Andrea Reider
Illustrations: DJ Simison, Carl Yoshihara, Rachel Gage
Cover: Bay Graphics

Printed on Recycled Paper

Order number DS43704
ISBN 1-57232- 468-6
5 6 7 8 9 10-ML-00 99

TERC

INVESTIGATIONS IN NUMBER, DATA, AND SPACE®

Principal Investigator **Susan Jo Russell**

Co-Principal Investigator **Cornelia C. Tierney**

Director of Research and Evaluation **Jan Mokros**

Director of K–2 Curriculum **Karen Economopoulos**

Curriculum Development

Karen Economopoulos
Marlene Kliman
Jan Mokros
Megan Murray
Susan Jo Russell
Tracey Wright

Evaluation and Assessment

Mary Berle-Carman
Jan Mokros
Andee Rubin

Teacher Support

Irene Baker
Megan Murray
Judy Storeygard
Tracey Wright

Technology Development

Michael T. Battista
Douglas H. Clements
Julie Sarama

Video Production

David A. Smith
Judy Storeygard

Administration and Production

Irene Baker
Amy Catlin

Cooperating Classrooms for This Unit

Barbara Reid
Boston Public Schools
Boston, MA

Joe Reilly
Malia Scott
Brookline Public Schools
Brookline, MA

Consultants and Advisors

Deborah Lowenberg Ball
Michael T. Battista
Marilyn Burns
Douglas H. Clements
Ann Grady

CONTENTS

TEACHER NOTES

WHERE TO START

The first-time user of *Survey Questions and Secret Rules* should read the following:

When you next teach this same unit, you can begin to read more of the background. Each time you present the unit, you will learn more about how your students understand the mathematical ideas.

ABOUT THE *INVESTIGATIONS* CURRICULUM

Investigations in Number, Data, and Space® is a K–5 mathematics curriculum with four major goals:

- to offer students meaningful mathematical problems
- to emphasize depth in mathematical thinking rather than superficial exposure to a series of fragmented topics
- to communicate mathematics content and pedagogy to teachers
- to substantially expand the pool of mathematically literate students

The *Investigations* curriculum embodies a new approach based on years of research about how children learn mathematics. Each grade level consists of a set of separate units, each offering 2–8 weeks of work. These units of study are presented through investigations that involve students in the exploration of major mathematical ideas.

Approaching the mathematics content through investigations helps students develop flexibility and confidence in approaching problems, fluency in using mathematical skills and tools to solve problems, and proficiency in evaluating their solutions. Students also build a repertoire of ways to communicate about their mathematical thinking, while their enjoyment and appreciation of mathematics grows.

The investigations are carefully designed to invite all students into mathematics—girls and boys, members of diverse cultural, ethnic, and language groups, and students with different strengths and interests. Problem contexts often call on students to share experiences from their family, culture, or community. The curriculum eliminates barriers—such as work in isolation from peers, or emphasis on speed and memorization—that exclude some students from participating successfully in mathematics. The following aspects of the curriculum ensure that all students are included in significant mathematics learning:

- Students spend time exploring problems in depth.
- They find more than one solution to many of the problems they work on.
- They invent their own strategies and approaches, rather than relying on memorized procedures.
- They choose from a variety of concrete materials and appropriate technology, including calculators, as a natural part of their everyday mathematical work.
- They express their mathematical thinking through drawing, writing, and talking.
- They work in a variety of groupings—as a whole class, individually, in pairs, and in small groups.
- They move around the classroom as they explore the mathematics in their environment and talk with their peers.

While reading and other language activities are typically given a great deal of time and emphasis in elementary classrooms, mathematics often does not get the time it needs. If students are to experience mathematics in depth, they must have enough time to become engaged in real mathematical problems. We believe that a minimum of 5 hours of mathematics classroom time a week—about an hour a day—is critical at the elementary level. The plan and pacing of the *Investigations* curriculum is based on that belief.

We explain more about the pedagogy and principles that underlie these investigations in Teacher Notes throughout the units. For correlations of the curriculum to the NCTM Standards and further help in using this research-based program for teaching mathematics, see the following books:

- *Implementing the* Investigations in Number, Data, and Space® *Curriculum*
- *Beyond Arithmetic: Changing Mathematics in the Elementary Classroom* by Jan Mokros, Susan Jo Russell, and Karen Economopoulos

This book is one of the curriculum units for *Investigations in Number, Data, and Space.* In addition to providing part of a complete mathematics curriculum for your students, this unit offers information to support your own professional development. You, the teacher, are the person who will make this curriculum come alive in the classroom; the book for each unit is your main support system.

Although the curriculum does not include student textbooks, reproducible sheets for student work are provided in the unit and are also available as Student Activity Booklets. Students work actively with objects and experiences in their own environment and with a variety of manipulative materials and technology, rather than with a book of instruction and problems. We strongly recommend use of the overhead projector as a way to present problems, to focus group discussion, and to help students share ideas and strategies.

Ultimately, every teacher will use these investigations in ways that make sense for his or her particular style, the particular group of students, and the constraints and supports of a particular school environment. Each unit offers information and guidance for a wide variety of situations, drawn from our collaborations with many teachers and students over many years. Our goal in this book is to help you, a professional educator, implement this curriculum in a way that will give all your students access to mathematical power.

Investigation Format

The opening two pages of each investigation help you get ready for the work that follows.

What Happens This gives a synopsis of each session or block of sessions.

Mathematical Emphasis This lists the most important ideas and processes students will encounter in this investigation.

What to Plan Ahead of Time These lists alert you to materials to gather, sheets to duplicate, transparencies to make, and anything else you need to do before starting.

INVESTIGATION 1

Sorting

What Happens

Sessions 1 and 2: Sorting Shapes Students explore attribute blocks, brainstorming words that describe the different attributes (their color, size, shape, and thickness). They then play Guess My Rule with Shapes, trying to guess the secret sorting rule. Finally, students sort paper Attribute Shape Cards, gluing them down to make a representation of the sorted shapes.

Session 3: Sorting After reading the book *Math Counts: Sorting* (if available), students discuss sorting objects by different attributes. In a game of Guess My Rule, the teacher sorts the students themselves according to secret rules based on their visible attributes. To prepare for upcoming sessions, students independently explore the class collections of buttons and lids.

Session 4: Sorting Buttons Students discuss the attributes of buttons, reading together *The Button Box* (if available). They then play Guess My Rule with Buttons.

Session 5: Sorting Lids As a class, students compare two lids, describing how they are the same and how they are different. This leads to the game Guess My Rule with Lids, with more secret sorting rules.

Session 6: Representations of Sorted Objects In an assessment activity, students sort buttons or lids and create a representation of their sorted groups. The sorting rule is recorded on a card that is fastened, facedown, to the representation, so that this student work can be used for further games of Guess My Rule.

Routines Refer to the section About Classroom Routines (pp. 100–107) for suggestions on integrating into the school day regular practice of mathematical skills in counting, exploring data, and understanding time and changes.

Mathematical Emphasis

- Identifying and describing attributes of various materials
- Using an attribute as a basis for sorting and categorizing a variety of objects
- Developing strategies to guess someone else's sorting rule
- Creating representations of sorted sets of objects

INVESTIGATION 1

What to Plan Ahead of Time

Materials

- Attribute blocks: class set of 60 (Sessions 1–2)
- Attribute Shape Cards: 30 per pair; use manufactured sets, or make your own (Sessions 1–2)
- String, cut into two-foot lengths and tied in loops (all sessions, optional)
- Chart paper (all sessions)
- Large paper (11 by 17 inches): 2 sheets per student, plus extras (Sessions 1, 2, and 6)
- Glue sticks and clear tape, to share (Sessions 1, 2, and 6)
- Paper scraps about 3 by 5 inches, 2 per student, plus extras. (Sessions 1, 2, and 6)
- Paper clips: 2 per student and extras (Sessions 1, 2, and 6)
- Buttons: at least 200 (Sessions 3, 4, and 6)
- Other collections, such as shells, rocks, hardware (all sessions, optional)
- *Math Counts: Sorting* by Henry Pluckrose (Childrens Press, 1995) (Session 3, optional)
- *The Button Box* by Margarette S. Reid (Dutton Children's Books, 1990) (Session 4, optional)

Other Preparation

- If you do not have manufactured Attribute Shape Cards, duplicate the master (p. 114) on red, yellow, and blue construction paper and cut apart to make a set of 30 cards for each pair of students. (Sessions 1–2)
- Assemble a large collection of lids for sorting, at least 15 lids per pair. Ask for discarded lids from home and check the lunch room. Look for a variety of lid types (metal, plastic, screw-on, snap-on). Avoid any with sharp edges. (Session 3)
- From your button collection, put together a set of about 20 buttons with *no* exact matches. (Session 4)
- Put together a set of about 20 lids with *no* exact matches. Also choose two lids that are alike in some ways, different in others. (Session 5)
- Duplicate the following student sheets and teaching resources, located at the end of this unit. If you have Student Activity Booklets, copy only items marked with an asterisk.

  Session 1
  Family letter* (p. 110): 1 per student. Sign and date it before duplicating. (You might also request donations of lids and buttons.)

  Session 3
  Student Sheet 1, Describe a Button (p. 111): 1 per student, homework

  Session 4
  Student Sheet 2, Sorting Things (p. 112): 1 per student, homework

  Session 5
  Student Sheet 3, Design a Lid (p. 113): 1 per student, homework
- If you plan to provide folders in which students will save their work for the entire unit, prepare these for distribution.

Sessions Within an investigation, the activities are organized by class session, a session being at least a one-hour math class. Sessions are numbered consecutively through an investigation. Often several sessions are grouped together, presenting a block of activities with a single major focus.

When you find a block of sessions presented together—for example, Sessions 1, 2, and 3—read through the entire block first to understand the overall flow and sequence of the activities. Make some preliminary decisions about how you will divide the activities into three sessions for your class, based on what you know about your students. You may need to modify your initial plans as you progress through the activities, and you may want to make notes in the margins of the pages as reminders for the next time you use the unit.

Be sure to read the Session Follow-Up section at the end of the session block to see what homework assignments and extensions are suggested as you make your initial plans.

While you may be used to a curriculum that tells you exactly what each class session should cover, we have found that the teacher is in a better position to make these decisions. Each unit is flexible and may be handled somewhat differently by every teacher. While we provide guidance for how many sessions a particular group of activities is likely to need, we want you to be active in determining an appropriate pace and the best transition points for your class. It is not unusual for a teacher to spend more or less time than is proposed for the activities.

Activities The activities include pair and small-group work, individual tasks, and whole-class discussions. In any case, students are seated together, talking and sharing ideas during all work times. Students most often work cooperatively, although each student may record work individually.

Choice Time In most units, some sessions are structured with activity choices. In these cases, students may work simultaneously on different activities focused on the same mathematical ideas. Students choose which activities they want to do, and they cycle through them. You will need to decide how to set up and introduce these activities and how to let students make their choices. Some teachers set up choices as stations around the room, while others post the list of available choices and allow students to collect their own materials and choose their own work space. You may need to experiment with a few different structures before finding a setup that works best for you.

Sessions 1 and 2

Sorting Shapes

Materials

- Attribute blocks (class set of 60)
- Chart paper
- Attribute Shape Cards (set of 30 per pair)
- String loops (1 per pair, optional)
- Large paper (1 sheet per student)
- Paper scraps about 3 by 6 inches (1 per student)
- Glue sticks, clear tape to share
- Paper clips (1 per student)

What Happens

Students explore attribute blocks, brainstorming words that describe the different attributes (their color, size, shape, and thickness). They then play Guess My Rule with Shapes, trying to guess the secret sorting rule. Finally, students sort paper Attribute Shape Cards, gluing them down to make a representation of the sorted shapes. Their work focuses on:

- exploring and describing attributes of shapes
- sorting shapes so that others can guess the sorting rule
- figuring out which attribute a set of blocks has been sorted by
- creating a representation and rule for a set of sorted blocks

Activity

Exploring Attribute Blocks

Introduce this unit by explaining that as part of their work in mathematics, the students are going to be collecting and sorting data. If you have presented the *Investigations* unit *Mathematical Thinking at Grade 1,* remind the class of the data they collected about themselves, through quick surveys.

When we collect a lot of data, one of the ways we make sense of it is by sorting it. When we sort, we look at how things are alike and how things are different. For example, maybe you've helped someone in your family sort the laundry, into "whites" and "darks." All the white or very light clothes go in one pile to be washed; anything darker goes in another pile.

This is the first thing we'll be sorting in class: colored shapes. *[Show the class set of attribute blocks.]* **We're going to be playing a game with these blocks, but first you're going to have some time to explore them. As you work with them, think about the different shapes, colors, and sizes.**

Extensions These follow-up activities are opportunities for some or all students to explore a topic in greater depth or in a different context. They are not designed for "fast" students; mathematics is a multifaceted discipline, and different students will want to go further in different investigations. Look for and encourage the sparks of interest and enthusiasm you see in your students, and use the extensions to help them pursue these interests.

Excursions Some of the *Investigations* units include excursions—blocks of activities that could be omitted without harming the integrity of the unit. This is one way of dealing with the great depth and variety of elementary mathematics—much more than a class has time to explore in any one year. Excursions give you the flexibility to make different choices from year to year, doing the excursion in one unit this time, and next year trying another excursion.

Tips for the Linguistically Diverse Classroom At strategic points in each unit, you will find concrete suggestions for simple modifications of the teaching strategies to encourage the participation of all students. Many of these tips offer alternative ways to elicit critical thinking from students at varying levels of English proficiency, as well as from other students who find it difficult to verbalize their thinking.

The tips are supported by suggestions for specific vocabulary work to help ensure that all students can participate fully in the investigations. The Preview for the Linguistically Diverse Classroom (p. I-20) lists important words that are assumed as part of the working vocabulary of the unit. Second-language learners will need to become familiar with these words in order to understand the problems and activities they will be doing. These terms can be incorporated into students' second-language work before or during the unit. Activities that can be used to present the words are found in the appendix, Vocabulary Support for Second-Language Learners (p. 108). In addition, ideas for making connections to students' language and cultures, included on the Preview page, help the class explore the unit's concepts from a multicultural perspective.

Classroom Routines Activities in counting, exploring data, and understanding time and changes are suggested for routines in the grade 1 *Investigations* curriculum. Routines offer ongoing work with this important content as a regular part of the school day. Some routines provide more practice with content presented in the curriculum; others extend the curriculum; still others explore new content areas.

Plan to incorporate a few of the routine activities into a standard part of your daily schedule, such as morning meeting. When opportunities arise, you can also include routines as part of your work in other subject areas (for example, keeping a weather chart for science). Most routines are short and can be done whenever you have a spare 10–15 minutes, such as before lunch or recess or at the end of the day.

You will need to decide how often to present routines, what variations are appropriate for your class, and at what points in the day or week you will include them. A reminder about classroom routines is included on the first page of each investigation. Whatever routines you choose, your students will gain the most from these routines if they work with them regularly.

Materials

A complete list of the materials needed for teaching this unit is found on p. I-17. Some of these materials are available in kits for the *Investigations* curriculum. Individual items can also be purchased from school supply dealers.

Classroom Materials In an active mathematics classroom, certain basic materials should be available at all times: interlocking cubes, pencils, unlined paper, graph paper, calculators, and things to count with. Some activities in this curriculum require scissors and glue sticks or tape. Stick-on notes and large paper are also useful materials throughout.

So that students can independently get what they need at any time, they should know where these materials are kept, how they are stored, and how they are to be returned to the storage area. Many teachers have found that stopping 5 minutes before the end of each session so that students can finish their work and clean up is helpful in maintaining classroom materials. You'll find that establishing such routines at the beginning of the year is well worth the time and effort.

Technology Calculators are introduced to students in the first unit of the grade 1 sequence, *Mathematical Thinking at Grade 1.* By freely exploring and experimenting, students become familiar with this important mathematical tool.

Computer activities at grade 1 use a software program, called Shapes, that was developed especially for the *Investigations* curriculum. This program is introduced in the geometry unit, *Quilt Squares and Block Towns.* Using *Shapes,* students explore two-dimensional geometry while making pictures and designs with pattern block shapes and tangram pieces.

Although the software is linked to activities only in the geometry unit, we recommend that students use it throughout the year. Thus, you may want to introduce it when you introduce pattern blocks in *Mathematical Thinking at Grade 1.* How you use the computer activities depends on the number of computers you have available. Suggestions are offered in the geometry unit for how to organize different types of computer environments.

Children's Literature Each unit offers a list of suggested children's literature (p. I-17) that can be used to support the mathematical ideas in the unit. Sometimes an activity is based on a specific children's book, with suggestions for substitutions where practical. While such activities can be adapted and taught without the book, the literature offers a rich introduction and should be used whenever possible.

Student Sheets and Teaching Resources Student recording sheets and other teaching tools needed for both class and homework are provided as reproducible blackline masters at the end of each unit. They are also available as Student Activity

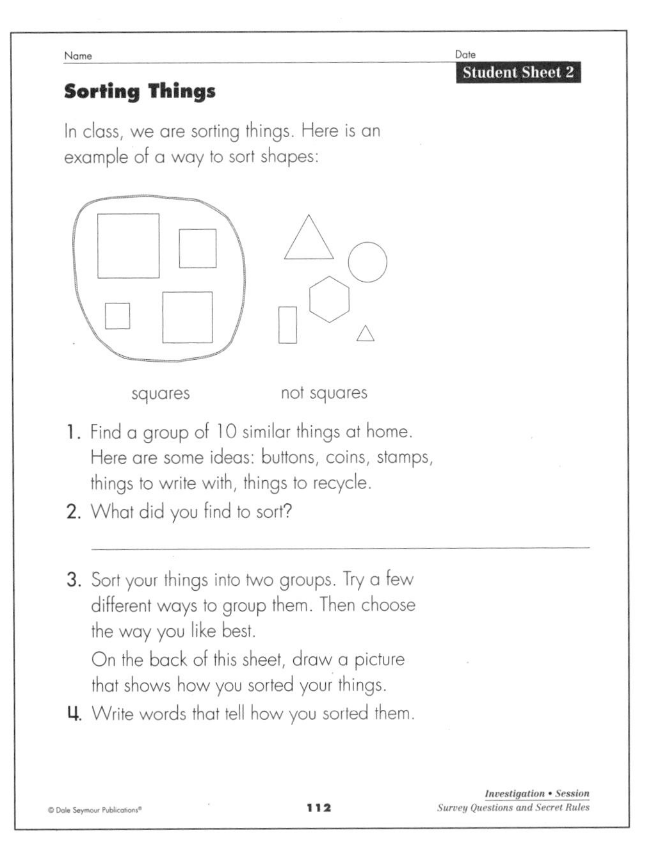
Name Date

Student Sheet 2

Sorting Things

In class, we are sorting things. Here is an example of a way to sort shapes:

1. Find a group of 10 similar things at home. Here are some ideas: buttons, coins, stamps, things to write with, things to recycle.
2. What did you find to sort?
3. Sort your things into two groups. Try a few different ways to group them. Then choose the way you like best. On the back of this sheet, draw a picture that shows how you sorted your things.
4. Write words that tell how you sorted them.

© Dale Seymour Publications® 112 *Investigation • Session Survey Questions and Secret Rules*

Booklets. These booklets contain all the sheets each student will need for individual work, freeing you from extensive copying (although you may need or want to copy the occasional teaching resource on transparency film or card stock, or make extra copies of a student sheet).

We think it's important that students find their own ways of organizing and recording their work. They need to learn how to explain their thinking with both drawings and written words, and how to organize their results so someone else can understand them. For this reason, we deliberately do not provide student sheets for every activity. Regardless of the form in which students do their work, we recommend that they keep a mathematics notebook or folder so that their work is always available for reference.

Homework In *Investigations,* homework is an extension of classroom work. Sometimes it offers review and practice of work done in class, sometimes preparation for upcoming activities, and sometimes numerical practice that revisits work in earlier units. Homework plays a role both in supporting students' learning and in helping inform families about the ways in which students in this curriculum work with mathematical ideas.

Depending on your school's homework policies and your own judgment, you may want to assign more homework than is suggested in the units. For this purpose you might use the practice pages, included as blackline masters at the end of this unit, to give students additional work with numbers.

For some homework assignments, you will want to adapt the activity to meet the needs of a variety of students in your class: those with special needs, those ready for more challenge, and second-language learners. You might change the numbers in a problem, make the activity more or less complex, or go through a sample activity with those who need extra help. You can modify any student sheet for either homework or class use. In particular, making numbers in a problem smaller or larger can make the same basic activity appropriate for a wider range of students.

Another issue to consider is how to handle the homework that students bring back to class—how to recognize the work they have done at home without spending too much time on it. Some teachers hold a short group discussion of different approaches to the assignment; others ask students to share and discuss their work with a neighbor, or post the homework around the room and give students time to tour it briefly. If you want to keep track of homework students bring in, be sure it ends up in a designated place.

***Investigations* at Home** It is a good idea to make your policy on homework explicit to both students and their families when you begin teaching with *Investigations.* How frequently will you be assigning homework? When do you expect homework to be completed and brought back to school? What are your goals in assigning homework? How independent should families expect their children to be? What should the parent or guardian's role be? The more explicit you can be about your expectations, the better the homework experience will be for everyone.

Investigations at Home (a booklet available separately for each unit, to send home with students) gives you a way to communicate with families about the work students are doing in class. This booklet includes a brief description of every session, a list of the mathematics content emphasized in each investigation, and a discussion of each homework assignment to help families more effectively support their children. Whether or not you are using the *Investigations* at Home booklets, we expect you to make your own choices about homework assignments. Feel free to omit any and to add extra ones you think are appropriate.

Family Letter A letter that you can send home to students' families is included with the blackline masters for each unit. Families need to be informed about the mathematics work in your classroom; they should be encouraged to participate in and support their children's work. A reminder to send home the letter for each unit appears in one of the early investigations. These letters are also available separately in Spanish, Vietnamese, Cantonese, Hmong, and Cambodian.

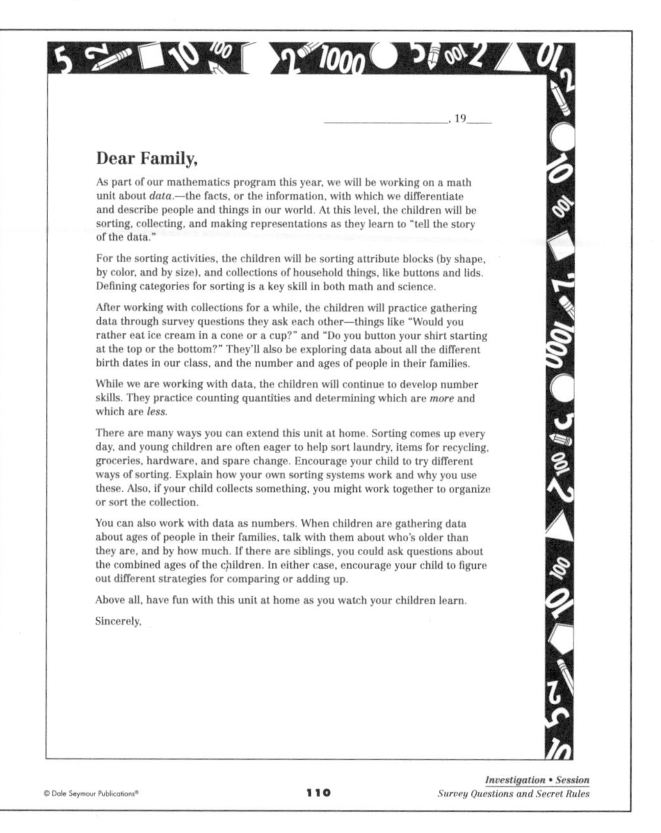

________________, 19____

Dear Family,

As part of our mathematics program this year, we will be working on a math unit about *data.*—the facts, or the information, with which we differentiate and describe people and things in our world. At this level, the children will be sorting, collecting, and making representations as they learn to "tell the story of the data."

For the sorting activities, the children will be sorting attribute blocks (by shape, by color, and by size), and collections of household things, like buttons and lids. Defining categories for sorting is a key skill in both math and science.

After working with collections for a while, the children will practice gathering data through survey questions they ask each other—things like "Would you rather eat ice cream in a cone or a cup?" and "Do you button your shirt starting at the top or the bottom?" They'll also be exploring data about all the different birth dates in our class, and the number and ages of people in their families.

While we are working with data, the children will continue to develop number skills. They practice counting quantities and determining which are *more* and which are *less.*

There are many ways you can extend this unit at home. Sorting comes up every day, and young children are often eager to help sort laundry, items for recycling, groceries, hardware, and spare change. Encourage your child to try different ways of sorting. Explain how your own sorting systems work and why you use these. Also, if your child collects something, you might work together to organize or sort the collection.

You can also work with data as numbers. When children are gathering data about ages of people in their families, talk with them about who's older than they are, and by how much. If there are siblings, you could ask questions about the combined ages of the children. In either case, encourage your child to figure out different strategies for comparing or adding up.

Above all, have fun with this unit at home as you watch your children learn.

Sincerely,

© Dale Seymour Publications® 110 *Investigation • Session* *Survey Questions and Secret Rules*

Help for You, the Teacher

Because we believe strongly that a new curriculum must help teachers think in new ways about mathematics and about their students' mathematical thinking processes, we have included a great deal of material to help you learn more about both.

About the Mathematics in This Unit This introductory section (p. I-18) summarizes the critical information about the mathematics you will be teaching. It describes the unit's central mathematical ideas and how students will encounter them through the unit's activities.

Teacher Notes These reference notes provide practical information about the mathematics you are teaching and about our experience with how students learn. Many of the notes were written in response to actual questions from teachers, or to discuss important things we saw happening in the field-test classrooms. Some teachers like to read them all before starting the unit, then review them as they come up in particular investigations.

Dialogue Boxes Sample dialogues demonstrate how students typically express their mathematical ideas, what issues and confusions arise in their thinking, and how some teachers have guided class discussions. These dialogues are based on the extensive classroom testing of this curriculum; many are word-for-word transcriptions of recorded class discussions. They are not always easy reading; sometimes it may take some effort to unravel what the students are trying to say. But this is the value of these dialogues; they offer good clues to how your students may develop and express their approaches and strategies, helping you prepare for your own class discussions.

Where to Start You may not have time to read everything the first time you use this unit. As a first-time user, you will likely focus on understanding the activities and working them out with your students. Read completely through each investigation before starting to present it. Also read those sections listed in the Contents (p. vi) under the heading Where to Start.

Teacher Note ***Sorting into Groups***

As students are sorting their buttons and lids, notice how they create categories. Some students may sort lids by color, putting the white ones in one group and "the rest" in another group. Some may spontaneously sort into more than two groups (e.g., WHITE lids, BLACK lids, BLUE lids, and the rest).

Students often find it easier to guess the teacher's rule than to come up with rules of their own. Many first graders will not use one rule for sorting the whole data set. For example, they may put BIG buttons in one group and RED buttons in another group. Some students may begin by looking for exact matches. As first graders spend more time exploring materials and talking about their features (or attributes), they will become familiar with a variety of sorting rules.

With "real-world" data sets, such as buttons or lids, there are many different attributes to be identified. Further, many of the attributes are "fuzzy"; that is, they are not as easy to define as those of the attribute blocks. There is no disagreement about whether an attribute block is RED or BLUE, but there is much room for discussion about which lids are SMALL or which buttons HAVE HOLES (do the ones on the back of shank buttons count?).

Some of the rules students have invented include these:

- LIDS THAT SOUND LIKE A DOORBELL
- LIDS THAT ROLL
- MY FAVORITE BUTTONS
- DARK COLORED BUTTONS

Rules like these raise legitimate problems of definition and clarification. Some of these issues can stimulate deeper thinking about classification and need not be simplified for students. However, students may need reminding that, just as when sorting people, they must sort in a way that is visually obvious to someone else. Thus, MY FAVORITES is not a rule that someone else can see.

Several particular issues often arise as students of this age work on sorting, all related to the problem of defining "fuzzy" categories.

Sometimes, a category that can be seen is still very difficult to describe. For example, one pair of students chose the rule LIDS YOU CAN ROLL. They demonstrated how to put your finger on the edge of the lid and flip it up so that it could roll on its edge. They had constructed a legitimate "fuzzy" characteristic, hard to define yet real. It was not clear if they meant LIDS THAT COULD ROLL or LIDS THAT COULD FLIP UP.

It is very tempting to "help" students with such an idea by getting them to substitute a simpler rule, one that is not so hard to define. A better approach is to help students clarify their rule simply by asking them which objects do and do not belong in their category, and why.

Some descriptions (like those of size) require a basis for comparison. For example, some students chose LIDS THAT ARE SMALL. When they began to field other students' guesses, they found they could not really agree on which lids did and did not fit their rule. When this happens, the teacher's role is to help students clarify their rule as suggested above. Before they challenge the rest of the class, ask them to decide which objects fit their rule, which objects don't fit, and why. Challenge their choices and help them verbalize their reasons. Even a measurement that seems exact may require further definition.

Another problem for first graders is that different words can be used to describe the same characteristic. The pair whose rule was BUTTONS THAT ARE SHINY did not accept student guesses of REFLECTING BUTTONS, METALLIC BUTTONS, or GOLD AND SILVER BUTTONS. There was outraged protest when they announced their rule because other students felt they had already identified the category in different words.

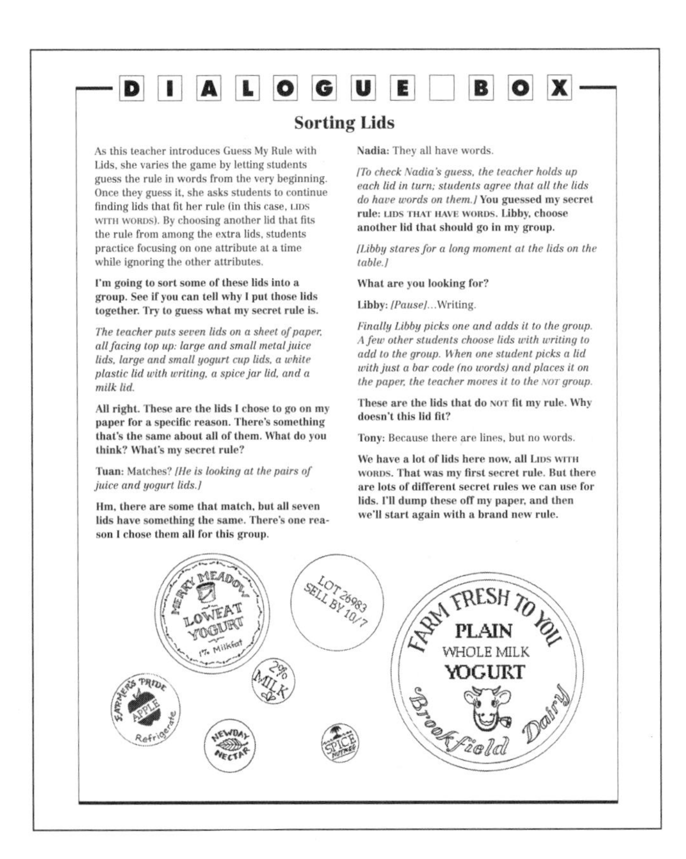

DIALOGUE BOX

Sorting Lids

As this teacher introduces Guess My Rule with Lids, she varies the game by letting students guess the rule in words from the very beginning. Once they guess it, she asks students to continue finding lids that fit her rule (in this case, LIDS WITH WORDS). By choosing another lid that fits the rule from among the extra lids, students practice focusing on one attribute at a time while ignoring the other attributes.

I'm going to sort some of these lids into a group. See if you can tell why I put those lids together. Try to guess what my secret rule is.

The teacher puts seven lids on a sheet of paper, all facing top up: large and small metal juice lids, large and small yogurt cup lids, a white plastic lid with writing, a spice jar lid, and a milk lid.

All right. These are the lids I chose to go on my paper for a specific reason. There's something that's the same about all of them. What do you think? What's my secret rule?

Tuan: Matches? *[He is looking at the pairs of juice and yogurt lids.]*

Hm, there are some that match, but all seven lids have something the same. There's one reason I chose them all for this group.

Nadia: They all have words.

[To check Nadia's guess, the teacher holds up each lid in turn; students agree that all the lids do have words on them.] **You guessed my secret rule: LIDS THAT HAVE WORDS. Libby, choose another lid that should go in my group.**

[Libby stares for a long moment at the lids on the table.]

What are you looking for?

Libby: *[Pause]*...Writing.

Finally Libby picks one and adds it to the group. A few other students choose lids with writing to add to the group. When one student picks a lid with just a bar code (no words) and places it on the paper, the teacher moves it to the NOT group.

These are the lids that do NOT fit my rule. Why doesn't this lid fit?

Tony: Because there are lines, but no words.

We have a lot of lids here now, all LIDS WITH WORDS. That was my first secret rule. But there are lots of different secret rules we can use for lids. I'll dump these off my paper, and then we'll start again with a brand new rule.

The *Investigations* curriculum incorporates the use of two forms of technology in the classroom: calculators and computers. Calculators are assumed to be standard classroom materials, available for student use in any unit. Computers are explicitly linked to one or more units at each grade level; they are used with the unit on 2-D geometry unit at each grade, as well as with some of the units on measuring, data, and changes.

Using Calculators

In this curriculum, calculators are considered tools for doing mathematics, similar to pattern blocks or interlocking cubes. Just as with other tools, students must learn both *how* to use calculators correctly and *when* they are appropriate to use. This knowledge is crucial for daily life, as calculators are now a standard way of handling numerical operations, both at work and at home.

Using a calculator correctly is not a simple task; it depends on a good knowledge of the four operations and of the number system, so that students can select suitable calculations and also determine what a reasonable result would be. These skills are the basis of any work with numbers, whether or not a calculator is involved.

Unfortunately, calculators are often seen as tools to check computations with, as if other methods are somehow more fallible. Students need to understand that any computational method can be used to check any other; it's just as easy to make a mistake on the calculator as it is to make a mistake on paper or with mental arithmetic. Throughout this curriculum, we encourage students to solve computation problems in more than one way in order to double-check their accuracy. We present mental arithmetic, paper-and-pencil computation, and calculators as three possible approaches.

In this curriculum we also recognize that, despite their importance, calculators are not always appropriate in mathematics instruction. Like any tools, calculators are useful for some tasks, but not for others. You will need to make decisions about when to allow students access to calculators and when to ask that they solve problems without them, so that they can concentrate on other tools and skills. At times when calculators are or are not appropriate for a particular activity, we make specific recommendations. Help your students develop their own sense of which problems they can tackle with their own reasoning and which ones might be better solved with a combination of their own reasoning and the calculator.

Managing calculators in your classroom so that they are a tool, and not a distraction, requires some planning. When calculators are first introduced, students often want to use them for everything, even problems that can be solved quite simply by other methods. However, once the novelty wears off, students are just as interested in developing their own strategies, especially when these strategies are emphasized and valued in the classroom. Over time, students will come to recognize the ease and value of solving problems mentally, with paper and pencil, or with manipulatives, while also understanding the power of the calculator to facilitate work with larger numbers.

Experience shows that if calculators are available only occasionally, students become excited and distracted when they are permitted to use them. They focus on the tool rather than on the mathematics. In order to learn when calculators are appropriate and when they are not, students must have easy access to them and use them routinely in their work.

If you have a calculator for each student, and if you think your students can accept the responsibility, you might allow them to keep their calculators with the rest of their individual materials, at least for the first few weeks of school. Alternatively, you might store them in boxes on a shelf, number each calculator, and assign a corresponding number to each student. This system can give students a sense of ownership while also helping you keep track of the calculators.

Using Computers

Students can use computers to approach and visualize mathematical situations in new ways. The computer allows students to construct and manipulate geometric shapes, see objects move according to rules they specify, and turn, flip, and repeat a pattern.

This curriculum calls for computers in units where they are a particularly effective tool for learning mathematics content. One unit on 2-D geometry at each of the grades 3–5 includes a core of activities that rely on access to computers, either in the classroom or in a lab. Other units on geometry, measurement, data, and changes include computer activities, but can be taught without them. In these units, however, students' experience is greatly enhanced by computer use.

The following list outlines the recommended use of computers in this curriculum:

Grade 1
Unit: *Survey Questions and Secret Rules* (Collecting and Sorting Data)
Software: Tabletop, Jr.
Source: Broderbund

Unit: *Quilt Squares and Block Towns* (2-D and 3-D Geometry)
Software: *Shapes*
Source: provided with the unit

Grade 2
Unit: *Mathematical Thinking at Grade 2* (Introduction)
Software: *Shapes*
Source: provided with the unit

Unit: *Shapes, Halves, and Symmetry* (Geometry and Fractions)
Software: Shapes
Source: provided with the unit

Unit: *How Long? How Far?* (Measuring)
Software: *Geo-Logo*
Source: provided with the unit

Grade 3
Unit: *Flips, Turns, and Area* (2-D Geometry)
Software: *Tumbling Tetrominoes*
Source: provided with the unit

Unit: *Turtle Paths* (2-D Geometry)
Software: *Geo-Logo*
Source: provided with the unit

Grade 4
Unit: *Sunken Ships and Grid Patterns* (2-D Geometry)
Software: *Geo-Logo*
Source: provided with the unit

Grade 5
Unit: *Picturing Polygons* (2-D Geometry)
Software: *Geo-Logo*
Source: provided with the unit

Unit: *Patterns of Change* (Tables and Graphs)
Software: *Trips*
Source: provided with the unit

Unit: *Data: Kids, Cats, and Ads* (Statistics)
Software: Tabletop, Sr.
Source: Broderbund

The software provided with the *Investigations* units uses the power of the computer to help students explore mathematical ideas and relationships that cannot be explored in the same way with physical materials. With the *Shapes* (grades 1–2) and *Tumbling Tetrominoes* (grade 3) software, students explore symmetry, pattern, rotation and reflection, area, and characteristics of 2-D shapes. With the *Geo-Logo* software (grades 3–5), students investigate rotations and reflections, coordinate geometry, the properties of 2-D shapes, and angles. The *Trips* software (grade 5) is a mathematical exploration of motion in which students run experiments and interpret data presented in graphs and tables.

We suggest that students work in pairs on the computer; this not only maximizes computer resources but also encourages students to consult, monitor, and teach one another. Generally, more than two students at one computer find it difficult to share. Managing access to computers is an issue for every classroom. The curriculum gives you explicit support for setting up a system. The units are structured on the assumption that you have enough computers for half your students to work on the machines in pairs at one time. If you do not have access to that many computers, suggestions are made for structuring class time to use the unit with five to eight computers, or even with fewer than five.

Assessment plays a critical role in teaching and learning, and it is an integral part of the *Investigations* curriculum. For a teacher using these units, assessment is an ongoing process. You observe students' discussions and explanations of their strategies on a daily basis and examine their work as it evolves. While students are busy recording and representing their work, working on projects, sharing with partners, and playing mathematical games, you have many opportunities to observe their mathematical thinking. What you learn through observation guides your decisions about how to proceed. In any of the units, you will repeatedly consider questions like these:

- Do students come up with their own strategies for solving problems, or do they expect others to tell them what to do? What do their strategies reveal about their mathematical understanding?
- Do students understand that there are different strategies for solving problems? Do they articulate their strategies and try to understand other students' strategies?
- How effectively do students use materials as tools to help with their mathematical work?
- Do students have effective ideas for keeping track of and recording their work? Does keeping track of and recording their work seem difficult for them?

You will need to develop a comfortable and efficient system for recording and keeping track of your observations. Some teachers keep a clipboard handy and jot notes on a class list or on adhesive labels that are later transferred to student files. Others keep loose-leaf notebooks with a page for each student and make weekly notes about what they have observed in class.

Assessment Tools in the Unit

With the activities in each unit, you will find questions to guide your thinking while observing the students at work. You will also find two built-in assessment tools: Teacher Checkpoints and embedded Assessment activities.

Teacher Checkpoints The designated Teacher Checkpoints in each unit offer a time to "check in" with individual students, watch them at work, and ask questions that illuminate how they are thinking.

At first it may be hard to know what to look for, hard to know what kinds of questions to ask. Students may be reluctant to talk; they may not be accustomed to having the teacher ask them about their work, or they may not know how to explain their thinking. Two important ingredients of this process are asking students open-ended questions about their work and showing genuine interest in how they are approaching the task. When students see that you are interested in their thinking and are counting on them to come up with their own ways of solving problems, they may surprise you with the depth of their understanding.

Teacher Checkpoints also give you the chance to pause in the teaching sequence and reflect on how your class is doing overall. Think about whether you need to adjust your pacing: Are most students fluent with strategies for solving a particular kind of problem? Are they just starting to formulate good strategies? Or are they still struggling with how to start? Depending on what you see as the students work, you may want to spend more time on similar problems, change some of the problems to use smaller numbers, move quickly to more challenging material, modify subsequent activities for some students, work on particular ideas with a small group, or pair students who have good strategies with those who are having more difficulty.

Embedded Assessment Activities Assessment activities embedded in each unit will help you examine specific pieces of student work, figure out what it means, and provide feedback. From the students' point of view, these assessment activities are no different from any others. Each is a learning experience in and of itself, as well as an opportunity for you to gather evidence about students' mathematical understanding.

The embedded assessment activities sometimes involve writing and reflecting; at other times, a discussion or brief interaction between student and teacher; and in still other instances, the creation and explanation of a product. In most cases, the assessments require that students show what they did, write or talk about it, or do both. Having to explain how they worked through a problem helps students be more focused and clear in their mathematical thinking. It also helps them realize that doing mathematics is a process that may involve tentative starts, revising one's approach, taking different paths, and working through ideas.

Teachers often find the hardest part of assessment to be interpreting their students' work. We provide guidelines to help with that interpretation. If you have used a process approach to teaching writing, the assessment in *Investigations* will seem familiar. For many of the assessment activities, a Teacher Note provides examples of student work and a commentary on what it indicates about student thinking.

Documentation of Student Growth

To form an overall picture of mathematical progress, it is important to document each student's work in journals, notebooks, or portfolios. The choice is largely a matter of personal preference; some teachers have students keep a notebook or folder for each unit, while others prefer one mathematics notebook, or a portfolio of selected work for the entire year. The final activity in each *Investigations* unit, called Choosing Student Work to Save, helps you and the students select representative samples for a record of their work.

This kind of regular documentation helps you synthesize information about each student as a mathematical learner. From different pieces of evidence, you can put together the big picture. This synthesis will be invaluable in thinking about where to go next with a particular child, deciding where more work is needed, or explaining to parents (or other teachers) how a child is doing.

If you use portfolios, you need to collect a good balance of work, yet avoid being swamped with an overwhelming amount of paper. Following are some tips for effective portfolios:

- Collect a representative sample of work, including some pieces that students themselves select for inclusion in the portfolio. There should be just a few pieces for each unit, showing different kinds of work—some assignments that involve writing, as well as some that do not.
- If students do not date their work, do so yourself so that you can reconstruct the order in which pieces were done.
- Include your reflections on the work. When you are looking back over the whole year, such comments are reminders of what seemed especially interesting about a particular piece; they can also be helpful to other teachers and to parents. Older students should be encouraged to write their own reflections about their work.

Assessment Overview

There are two places to turn for a preview of the assessment opportunities in each *Investigations* unit. The Assessment Resources column in the unit Overview Chart (pp. I-13–I-16) identifies the Teacher Checkpoints and Assessment activities embedded in each investigation, guidelines for observing the students that appear within classroom activities, and any Teacher Notes and Dialogue Boxes that explain what to look for and what types of student responses you might expect to see in your classroom. Additionally, the section About the Assessment in This Unit (p. I-19) gives you a detailed list of questions for each investigation, keyed to the mathematical emphases, to help you observe student growth.

Depending on your situation, you may want to provide additional assessment opportunities. Most of the investigations lend themselves to more frequent assessment, simply by having students do more writing and recording while they are working.

Survey Questions and Secret Rules

Content of This Unit In *Survey Questions and Secret Rules,* students identify and describe attributes of objects, such as color, size, and shape. They sort and categorize attribute blocks, buttons, household lids, and even each other as they play a variety of Guess My Rule games. Students create representations of their object sets to demonstrate "secret rules" used for sorting. They also invent their own representations to show the results of class surveys on personal preferences. Students continue to collect data about themselves (birth dates, ages of classmates and family members), which they sort and represent in ways that help them compare several categories of data. Their work concludes with a comparison of two sets of class attendance data, for a usual and an unusual (imaginary) day.

Connections with Other Units If you are doing the full-year *Investigations* curriculum in the suggested sequence for grade 1, this is the third of six units. The work in this unit continues the exploration of data begun in *Mathematical Thinking at Grade 1;* in that unit, students also used surveys to collect data about themselves which they organized, represented, and described.

Investigations Curriculum ▪ Suggested Grade 1 Sequence

Mathematical Thinking at Grade 1 (Introduction)

Building Number Sense (The Number System)

▶ *Survey Questions and Secret Rules* (Collecting and Sorting Data)

Quilt Squares and Block Towns (2-D and 3-D Geometry)

Number Games and Story Problems (Addition and Subtraction)

Bigger, Taller, Heavier, Smaller (Measuring)

Investigation 1 ■ Sorting

Class Sessions	Activities	Pacing
Sessions 1 and 2 (p. 4) SORTING SHAPES	Exploring Attribute Blocks Guess My Rule with Shapes Guess My Partner's Rule with Shapes Teacher Checkpoint: Representations of Sorted Shapes Homework: Family Connection	minimum 2 hr
Session 3 (p. 14) SORTING	*Math Counts: Sorting* Guess My Rule Exploring Lids and Buttons Homework: Describe a Button	minimum 1 hr
Session 4 (p. 18) SORTING BUTTONS	*The Button Box* Guess My Rule with Buttons Guess My Partner's Rule with Buttons Homework: Sorting Things	minimum 1 hr
Session 5 (p. 24) SORTING LIDS	Comparing Lids Guess My Rule with Lids Guess My Partner's Rule with Lids Homework: Design a Lid Extension: Same and Different	minimum 1 hr
Session 6 (p. 29) REPRESENTATIONS OF SORTED OBJECTS	Assessment: A Favorite Sorting Rule	minimum 1 hr
Classroom Routines		

Mathematical Emphasis

- Identifying and describing attributes of various materials
- Using an attribute as a basis for sorting and categorizing a variety of objects
- Developing strategies to guess someone else's sorting rule
- Creating representations of sorted sets of objects

Assessment Resources

Teacher Checkpoint: Representations of Sorted Shapes (p. 9)

Keeping Track of Students' Work (Teacher Note, p. 11)

Observing the Students (pp. 10, 20, 26)

Assessment: A Favorite Sorting Rule (p. 29)

Materials

Attribute blocks
Attribute Shape Cards
Buttons
Household lids
String (optional)
Chart paper
Large paper (11 by 17 in.)
Glue sticks and tape
Paper clips
Math Counts: Sorting by H. Pluckrose (optional)
The Button Box by M. Reid (optional)
Student Sheets 1–3
Teaching resource sheets

Investigation 2 ■ Survey Questions

Class Sessions	Activities	Pacing
Sessions 1 and 2 (p. 34) WOULD YOU RATHER?	Would You Rather Be an Eagle or a Whale? Representations of Eagle and Whale Data Choosing a Question and Making a Plan Homework: Collections for Sorting Extension: More Eagles and Whales	minimum 2 hr
Sessions 3 and 4 (p. 43) MORE SURVEYS AND SORTING	Not-Boxes Introducing Choice Time Extension: Sorting with Tabletop, Jr.	minimum 2 hr
Sessions 5 and 6 (p. 48) REPRESENTING AND SHARING SURVEY RESULTS	Teacher Checkpoint: Making Representations of Our Data Our Findings Presenting to the Group Homework: Find Your Birth Date Extension: Other Surveys	minimum 2 hr
Classroom Routines		

Mathematical Emphasis

- Making a plan for gathering and recording data
- Sorting and categorizing data
- Inventing and constructing data representations
- Explaining and interpreting results of surveys
- Presenting data to others in a way that communicates information clearly

Assessment Resources

Observing the Students (pp. 37, 46, 49)

Teacher Checkpoint: Making Representations of Our Data (p. 48)

Materials

Interlocking cubes
Stick-on notes
Buttons
Household lids
Attribute blocks
Attribute Shape Cards
String (optional)
Large paper (11 by 17 in.)
Dot stickers, colored paper squares
Markers, glue
Clipboards (optional)
Class lists
Flat boxes
Student Sheets 4–6
Teaching resource sheets

Investigation 3 ■ Birthdays

Class Sessions	Activities	Pacing
Session 1 (p. 56) WHEN IS YOUR BIRTHDAY?	When Is Your Birthday? How Many in Each Month? Homework: Calendars	minimum 1 hr
Session 2 (p. 63) WHOSE BIRTHDAY COMES NEXT?	Organizing the Birthday Data Whose Birthday Comes Next? Birthday Lineup Extension: Special Celebrations	minimum 1 hr
Session 3 (Excursion)* (p. 70) TIMELINE	*A House for Hermit Crab* Creating a Timeline	minimum 1 hr
Classroom Routines		

* Excursions can be omitted without harming the integrity or continuity of the unit, but offer good mathematical work if you have time to include them.

Mathematical Emphasis

- Becoming familiar with calendar features
- Grouping and describing data about birthdays
- Ordering data about birthdays

Assessment Resources

Observing the Students (p. 64)

Materials

Index cards or half sheets of paper
Small envelopes
Crayons or markers
Drawing paper
Chart paper
A House for Hermit Crab by E. Carle
Calendars
Student Sheet 7
Teaching resource sheet

Investigation 4 ■ Ages and Attendance

Class Sessions	Activities	Pacing
Session 1 (p. 76) HOW OLD ARE WE?	How Old Are We? Listing Family Ages Homework: Collecting Family Age Data	minimum 1 hr
Sessions 2 and 3 (p. 80) REPRESENTING DATA ON FAMILY AGES	Family Portraits Ordering People by Age Who's the Mystery Person? Organizing Data: Us and Our Siblings Extension: Family Math Problems Extension: Ordering Family Members	minimum 1 hr
Sessions 4 and 5 (p. 90) ATTENDANCE COMPARISONS	Signing In A Most Unusual Day Assessment: Representing a Most Unusual Day Comparing Usual and Unusual Days Choosing Student Work to Save	minimum 1 hr
Classroom Routines		

Mathematical Emphasis

- Creating a variety of representations of several categories of data
- Describing data qualitatively and quantitatively
- Interpreting data that shows values (ages) and categories (siblings and selves) at the same time
- Comparing two data sets

Assessment Resources

Observing the Students (pp. 85, 94)

Assessment: Representing a Most Unusual Day (p. 94)

Assessment: Representing a Most Unusual Day (Teacher Note, p. 98)

Materials

Interlocking cubes
Crayons or markers
Scissors
Stick-on notes (two colors)
Chart paper
Construction paper
Large paper (11 by 17 in.)
Stick-on notes, dot stickers
Glue sticks
Class lists
Student Sheets 8–9
Teaching resource sheet

MATERIALS LIST

Following are the basic materials needed for the activities in this unit. Many items can be purchased from the publisher, either individually or in the Teacher Resource Package and the Student Materials Kit for grade 1. Detailed information is available on the *Investigations* order form. To obtain this form, call toll-free 1-800-872-1100 and ask for a Dale Seymour customer service representative.

Attribute blocks: class set of 60

Attribute Shape Cards (manufactured, or use blackline masters to make your own)

Buttons: about 200 assorted

Household lids: about 200 assorted

Interlocking cubes: about 500

Kid Pins and survey boards from *Mathematical Thinking at Grade 1* (optional)

Calendars, an assortment of current and old

Math Counts: Sorting by Henry Pluckrose (optional)

The Button Box by Margarette S. Reid (optional)

A House for Hermit Crab by Eric Carle (for Excursion session)

Clipboards, or make from heavy cardboard and binder clips (optional)

String loops (optional)

Flat boxes (3–6)

Crayons and markers

Index cards (5 by 8 inches) or half sheets of paper

Stick-on notes in two colors

Envelopes (1 per student, to hold data sets)

Large paper, 11 by 17 inches (4–5 sheets per student)

Drawing paper (13 sheets)

Construction paper

Chart paper

Scrap paper

Dot stickers, squares of colored paper, or similar materials for representing data (optional)

Paper clips

Scissors

Tape

Glue sticks

The following materials are provided at the end of this unit as blackline masters. A Student Activity Booklet containing all student sheets and teaching resoures needed for individual work is available.

Family Letter (p. 110)

Student Sheets 1–9 (p. 111)

Teaching Resources:
- Shape Cards (p. 114)
- Survey Questions (p. 118)
- Shape Rule Cards (p. 120)
- Birthday Grid (p. 122)
- 100 Chart (p. 125)

Practice Pages (p. 127)

Related Children's Literature

Baer, Edith. *This Is the Way We Eat Our Lunch.* New York: Scholastic, 1995.

Burningham, John. *Would You Rather....* New York: Thomas Crowell, 1978.

Carle, Eric. *A House for Hermit Crab.* New York: Scholastic, 1987.

Drescher, Joan. *The Birth-Order Blues.* New York: Viking, 1993.

Pluckrose, Henry. *Math Counts: Sorting.* Chicago: Childrens Press, 1995.

Reid, Margarette S. *The Button Box.* New York: Dutton Children's Books, 1990.

ABOUT THE MATHEMATICS IN THIS UNIT

It's a common conception that data are numbers—and indeed, they often are. But in *Investigations,* the exploration of data at grade 1 emphasizes categorical data, looking at things that can be categorized on the basis of a feature they have in common. For example, students categorize themselves according to the month in which they were born. In later grades, students will work with *numerical* data, or information they collect through measuring or counting. Through the activities in this unit, students work with categorical data in three important ways: they sort objects and data; they invent ways to represent data; and they learn to make sense of data representations.

Sorting Objects and Data The first question students explore is this: How can a group of objects be organized so that we can see how they are similar or different? Being able to look at an object (for example, a button) and attend primarily to one of its features (color, size, number of holes) is a key component of data literacy. Scientists and social scientists spend a great deal of time thinking about how to categorize data; their choice provides an important framework for analyzing results of a study. In this unit students develop categorization skills by making decisions about what categories to use as well as which data belong in which categories.

As students look carefully at attribute blocks, buttons, and lids, they describe the similarities and differences among objects in each collection, discovering that categorical data can be grouped in many ways. In the game Guess My Rule, students secretly identify one attribute of an object (such as *blue*) and group a set of objects that share this attribute, so that someone else can guess the common feature.

Inventing Representations of Data When asked to represent what they have found, students face two further questions: What information is important to represent? How can we arrange it so that someone else understands what we have found out? First graders communicate through drawings a great deal of information about what they know and think. In this unit, students create pictorial representations that communicate their mathematical discoveries during their work with shapes, buttons, and lids, as well as with data they collect through survey questions.

How can we make a visual representation to "tell a story" about, for example, the number of children in class who would rather be a whale for a day and the number who would rather be an eagle? Students find ways to distinguish between two categories of data and to communicate the relative sizes of the categories.

Making Sense of Data Representations When looking at their own or someone else's representations, students explore another question: How can we describe the story this data tells? Students are asked to make sense of a variety of representations—pictures, tallies, bar graphs—that help communicate information clearly to others. They are also asked to describe and make sense of their own and others' representations: pictures of sorted object sets, the class birthday data, ordered representations of the ages of people in their families, and two sets of attendance data.

Working with Number The connection between number and data is inherent throughout this unit. One reason that students categorize data is to compare the size of two categories. They compare numbers when they think about which group is bigger and how much bigger it is. They count, combine, and compare in order to keep track of their data, as well as to make statements about what they found out. Numbers constitute a powerful tool for visualizing and analyzing data that will be important in all students' experiences with data.

Mathematical Emphasis At the beginning of each investigation, the Mathematical Emphasis section tells you what is most important for students to learn about during that investigation. Many of these understandings and processes are difficult and complex. Students gradually learn more and more about each idea over many years of schooling. Individual students will begin and end the unit with different levels of knowledge and skill, but all will develop a sense of what it means to investigate a data question as they collect, sort, represent, and describe categorical data.

ABOUT THE ASSESSMENT IN THIS UNIT

Throughout the *Investigations* curriculum, there are many opportunities for ongoing daily assessment as you observe, listen to, and interact with students at work. In this unit you will find two Teacher Checkpoints:

Investigation 1, Session 1:
Representations of Sorted Shapes (p. 9)

Investigation 2, Sessions 5–6:
Making Representations of Our Data (p. 48)

This unit also has two embedded assessment activities:

Investigation 1, Session 6:
A Favorite Sorting Rule (p. 29)

Investigation 4, Sessions 4–5:
Representing a Most Unusual Day (p. 94)

In addition, you can use almost any activity in this unit to assess your students' needs and strengths. Listed below are questions to help you focus your observations in each investigation. You may want to keep track of your observations for each student to help you plan your curriculum and monitor students' growth. Suggestions for documenting student growth can be found in the section About Assessment (p. I-10).

Investigation 1: Sorting

- Are students identifying attributes to sort by? Are they distinguishing features that are alike and different?
- Can students choose a clear category and sort one set of objects according to their rule? If not, how are they grouping their objects? Are students comfortable sorting a small group of objects in more than one way?
- What guessing strategies are students using to guess someone else's sorting rule? Do they pay attention to both positive examples (people/objects that fit the rule) and negative examples (people/objects that don't fit)?
- Are students' representations clear to others? Do they use objects? words? both? Are their groups distinctive? Do students' written rules reflect the sorted groups on their page?

Investigation 2: Survey Questions

- Are students able to figure out a workable plan for gathering data? How are they keeping track?
- Are students grouping data that belong together? Have all the data been accounted for?
- What ideas are students using to make a representation that communicates the results of their surveys to others? What aspects need further clarification? Are their categories distinct?
- How are students explaining and interpreting survey results to you? to themselves? to other students?

Investigation 3: Birthdays

- How familiar are students with calendars, the months, and their birth dates?
- How are students grouping the class data about birthdays to show which cards belong together?
- Are students putting the data in order in consistent ways? What resources (such as calendars, or what they know about numbers) are they using as they order the data?

Investigation 4: Ages and Attendance

- How are students representing information about several categories of data? Are students accurately keeping track of their data?
- How are students describing data (qualitatively and quantitatively)?
- How are students making sense of data that shows values and categories at the same time?
- What strategies are students using to compare data sets? Are they making use of their own ordering of data?

PREVIEW FOR THE LINGUISTICALLY DIVERSE CLASSROOM

In the *Investigations* curriculum, mathematical vocabulary is introduced naturally during the activities. We don't ask students to learn definitions of new terms; rather, they come to understand such words as *triangle, add, compare, data,* and *graph* by hearing them used frequently in discussion as they investigate new concepts. This approach is compatible with current theories of second-language acquisition, which emphasize the use of new vocabulary in meaningful contexts while students are actively involved with objects, pictures, and physical movement.

Listed below are some key words used in this unit that will not be new to most English speakers at this age level, but may be unfamiliar to students with limited English proficiency. You will want to spend additional time working on these words with your students who are learning English. If your students are working with a second-language teacher, you might enlist your colleague's aid in familiarizing students with these words, before and during this unit. In the classroom, look for opportunities for students to hear and use these words. Activities you can use to present the words are given in the appendix, Vocabulary Support for Second-Language Learners (p. 108).

set, group In the first investigation, the students sort and classify objects into *sets* or *groups* defined by one attribute the objects share.

go together, similar, alike As they sort, students group together objects that are *similar* or *alike*—objects that *go together.*

month (and month names), order In the third investigation, students collect and organize data about their birthdays. While students are not expected to know all twelve months or their order (they will explore both during the investigation), they will benefit from some familiarity with the month names and the term *order.*

most, least As students look at the data they collect in the third and fourth investigations, they use these comparative terms as they talk about the data in different categories.

Multicultural Extensions for All Students

- For the sorting activities in Investigations 1 and 2, students might bring in lids from products from other countries. They also might work with collections of items like coins and stamps that represent different cultures.
- When you are discussing birthdays in Investigation 3, students might share the ways they celebrate birthdays (or other special days) in their culture. For example, a student who knows a birthday song in another language could teach it to the class. What games and foods are typical of birthday celebrations in other cultures?

Investigations

INVESTIGATION 1

Sorting

What Happens

Sessions 1 and 2: Sorting Shapes Students explore attribute blocks, brainstorming words that describe the different attributes (their color, size, shape, and thickness). They then play Guess My Rule with Shapes, trying to guess the secret sorting rule. Finally, students sort paper Attribute Shape Cards, gluing them down to make a representation of the sorted shapes.

Session 3: Sorting After reading the book *Math Counts: Sorting* (if available), students discuss sorting objects by different attributes. In a game of Guess My Rule, the teacher sorts the students themselves according to secret rules based on their visible attributes. To prepare for upcoming sessions, students independently explore the class collections of buttons and lids.

Session 4: Sorting Buttons Students discuss the attributes of buttons, reading together *The Button Box* (if available). They then play Guess My Rule with Buttons.

Session 5: Sorting Lids As a class, students compare two lids, describing how they are the same and how they are different. This leads to the game Guess My Rule with Lids, with more secret sorting rules.

Session 6: Representations of Sorted Objects In an assessment activity, students sort buttons or lids and create a representation of their sorted groups. The sorting rule is recorded on a card that is fastened, facedown, to the representation, so that this student work can be used for further games of Guess My Rule.

Routines Refer to the section About Classroom Routines (pp. 100–107) for suggestions on integrating into the school day regular practice of mathematical skills in counting, exploring data, and understanding time and changes.

Mathematical Emphasis

- Identifying and describing attributes of various materials
- Using an attribute as a basis for sorting and categorizing a variety of objects
- Developing strategies to guess someone else's sorting rule
- Creating representations of sorted sets of objects

What to Plan Ahead of Time

Materials

- Attribute blocks: class set of 60 (Sessions 1–2)
- Attribute Shape Cards: 30 per pair; use manufactured sets, or make your own (Sessions 1–2)
- String, cut into two-foot lengths and tied in loops (all sessions, optional)
- Chart paper (all sessions)
- Large paper (11 by 17 inches): 2 sheets per student, plus extras (Sessions 1, 2, and 6)
- Glue sticks and clear tape, to share (Sessions 1, 2, and 6)
- Paper scraps about 3 by 5 inches, 2 per student, plus extras. (Sessions 1, 2, and 6)
- Paper clips: 2 per student and extras (Sessions 1, 2, and 6)
- Buttons: at least 200 (Sessions 3, 4, and 6)
- Other collections, such as shells, rocks, hardware (all sessions, optional)
- *Math Counts: Sorting* by Henry Pluckrose (Childrens Press, 1995) (Session 3, optional)
- *The Button Box* by Margarette S. Reid (Dutton Children's Books, 1990) (Session 4, optional)

Other Preparation

- If you do not have manufactured Attribute Shape Cards, duplicate the master (p. 114) on red, yellow, and blue construction paper and cut apart to make a set of 30 cards for each pair of students. (Sessions 1–2)
- Assemble a large collection of lids for sorting, at least 15 lids per pair. Ask for discarded lids from home and check the lunch room. Look for a variety of lid types (metal, plastic, screw-on, snap-on). Avoid any with sharp edges. (Session 3)
- From your button collection, put together a set of about 20 buttons with *no* exact matches. (Session 4)
- Put together a set of about 20 lids with *no* exact matches. Also choose two lids that are alike in some ways, different in others. (Session 5)
- Duplicate the following student sheets and teaching resources, located at the end of this unit. If you have Student Activity Booklets, copy only items marked with an asterisk.

  Session 1
  Family letter* (p. 110): 1 per student. Sign and date it before duplicating. (You might also request donations of lids and buttons.)

  Session 3
  Student Sheet 1, Describe a Button (p. 111): 1 per student, homework

  Session 4
  Student Sheet 2, Sorting Things (p. 112): 1 per student, homework

  Session 5
  Student Sheet 3, Design a Lid (p. 113): 1 per student, homework

- If you plan to provide folders in which students will save their work for the entire unit, prepare these for distribution.

Sessions 1 and 2

Sorting Shapes

Materials

- Attribute blocks (class set of 60)
- Chart paper
- Attribute Shape Cards (set of 30 per pair)
- String loops (1 per pair, optional)
- Large paper (1 sheet per student)
- Paper scraps about 3 by 6 inches (1 per student)
- Glue sticks, clear tape to share
- Paper clips (1 per student)

What Happens

Students explore attribute blocks, brainstorming words that describe the different attributes (their color, size, shape, and thickness). They then play Guess My Rule with Shapes, trying to guess the secret sorting rule. Finally, students sort paper Attribute Shape Cards, gluing them down to make a representation of the sorted shapes. Their work focuses on:

- exploring and describing attributes of shapes
- sorting shapes so that others can guess the sorting rule
- figuring out which attribute a set of blocks has been sorted by
- creating a representation and rule for a set of sorted blocks

Activity

Exploring Attribute Blocks

Introduce this unit by explaining that as part of their work in mathematics, the students are going to be collecting and sorting data. If you have presented the *Investigations* unit *Mathematical Thinking at Grade 1,* remind the class of the data they collected about themselves, through quick surveys.

When we collect a lot of data, one of the ways we make sense of it is by sorting it. When we sort, we look at how things are alike and how things are different. For example, maybe you've helped someone in your family sort the laundry, into "whites" and "darks." All the white or very light clothes go in one pile to be washed; anything darker goes in another pile.

This is the first thing we'll be sorting in class: colored shapes. *[Show the class set of attribute blocks.]* **We're going to be playing a game with these blocks, but first you're going to have some time to explore them. As you work with them, think about the different shapes, colors, and sizes.**

Distribute handfuls of attribute blocks to pairs or small groups of students. Allow about 15 minutes for students to use the blocks in any way that makes sense to them. Watch what students do, and occasionally ask them what they are noticing about the blocks. At the end of the exploratory time, collect the blocks.

What did you notice about these blocks? Tell me something about them.

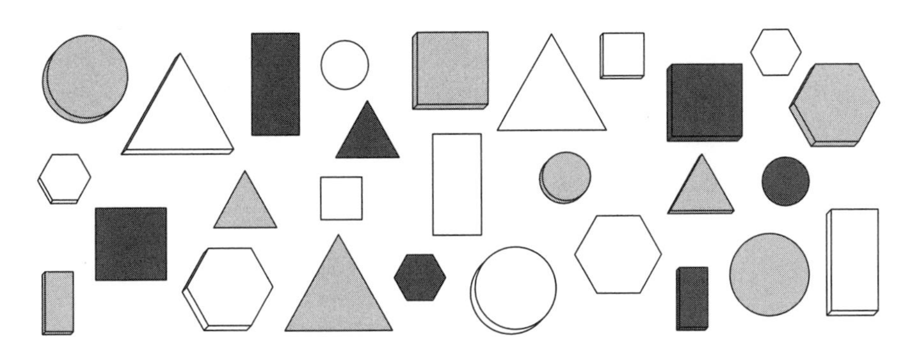

What different colors are there? How many different shapes are there? What are some things you noticed about the shapes? What would you say about the size of these blocks?

❖ **Tip for the Linguistically Diverse Classroom** During this discussion, have the attribute blocks available so that you or students may point to them as needed.

Keep track of students' ideas on chart paper. This is an opportunity to introduce or reinforce the names of the shapes (circle, square, triangle, rectangle, hexagon) as well as other terms related to shape, size, color, and grouping. See the **Dialogue Box,** What Did You Notice? (p. 12), for one such classroom conversation.

Activity

Guess My Rule with Shapes

For this game, students sit in a circle or semicircle on the floor. Start by placing 10–12 attribute blocks out where everyone can see them (in the center of the student circle or along the chalkboard ledge). Keep extra blocks nearby.

Tell the students to watch while you get ready to play a new game. Sort the blocks into two groups, according to one of the attributes students identified in the preceding discussion (listed on chart paper). For example, you might sort them by size (BIG and SMALL). If you want to define the groups more clearly, place one group on a sheet of paper or inside a loop of string.

I just sorted these shapes into two groups, using a secret rule. Look carefully at the two groups. If you can guess how I sorted them, don't say your guess out loud—we don't want to end the game too soon, before everyone has had a chance to figure out the secret rule.

Instead of calling out your guess, here's how we play: You try to guess my secret rule by silently putting another shape into one of my two groups.

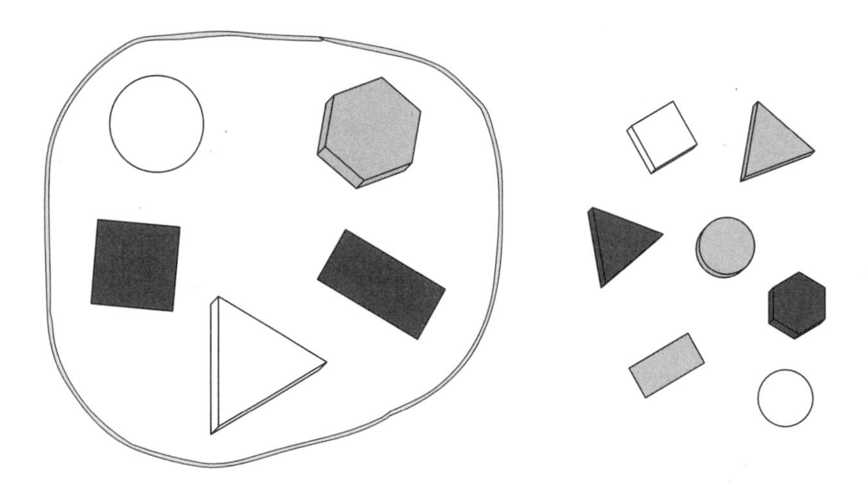

Give students a moment to study the blocks. When someone is ready to guess, call on that student to choose a shape and place it in one of the groups. After at least four or five blocks have been correctly placed, ask someone to say what they think the secret rule is. Students who agree raise their hand. If some students disagree, ask them to explain their thinking. If many students are excited about guessing the rule, they might whisper the rule altogether into the center of the circle.

Model writing the sorting rule on the board. Students sometimes have trouble putting their rules into words; help them find a form that they are comfortable with. For example, here are two possible ways of structuring a rule:

> My secret rule is BIG BLOCKS.
>
> I put together all blocks that are big.

For the next round of the game, sort the blocks another way according to a different attribute (by color or by shape). Explain that you have tried to sort the shapes in as clear a way as possible because you want everyone to be able to guess your rule; you are *not* trying to trick them. Repeat the guessing process, and finish by writing your rule on the board. The **Dialogue Box,** Guess My Rule with Shapes (p. 13), demonstrates how one teacher helped her students learn the game.

When the class seems comfortable with the game, invite a student to sort the shapes another way.

Let's have someone else make up a secret rule. Does anyone see some shapes that go together?

Choose a volunteer to arrange the shapes according to a new secret rule. (Decide whether the volunteer should first whisper the idea for a secret rule in your ear.) For example, a student might put all the SQUARES together, or all the YELLOW blocks. As before, other students try to guess the secret rule by adding blocks to the groups until everyone can guess the rule. When the rule has been guessed, write it on the board.

If time permits, a few more students may take a turn. At this point, allow them to create whatever arrangement makes sense to them, even if they do not use all the shapes or do not seem to make a consistent arrangement. If it is difficult to guess a particular rule, ask the student to talk through the sorting, to explain why he or she is putting certain shapes together.

This can be a good time to end Session 1, telling students that they will have more time with these blocks during the next session. If you end Session 1 here, you might begin Session 2 with a review of the Guess My Rule with Shapes game, leading the group through one or two rounds.

Activity

Guess My Partner's Rule with Shapes

Show students the Attribute Shape Cards in three colors. Explain that this is a paper version of the shape blocks they have been playing with, although there are no thick and thin shapes pictured.

You will have chance to sort these paper shapes with a partner. One of you thinks of a secret rule and puts some of the shapes in a group. Your partner then tries to guess your secret rule the same way we did in class: not by *saying* the rule, but by *showing* what the rule is. That is, your partner will try adding another shape to your group. Then, still without saying your rule, you will let your partner know if the added shape fits your rule or not.

When the rule has been guessed, you switch places. Now it is your partner's turn to think of a secret rule and sort the paper shapes.

Distribute a loop of string or a sheet of unlined paper and a set of Attribute Shape Cards to each pair of students. Students work in pairs, sorting and resorting in a variety of ways and guessing each other's rules. For practice with putting their rules in writing, you might leave possible frameworks on the board for students to copy:

My secret rule is ____________

I put together all ____________

As students work, circulate to observe how they are sorting and what information they use to guess each other's rules. Students generally find it easier to guess the teacher's rule than to create rules of their own. As they begin doing their own sorting, many first graders will not yet sort consistently according to one rule that they follow with all their shapes. It may help some student pairs to mutually decide on a rule, and then play by having one of them place shapes while the other gives information about whether the shapes fit the rule.

If you notice that students are sometimes changing their sorting rule midway through a round, remind them that they may choose only one way of sorting for each round. Some students may find it helpful to draw or write their rule on a slip of paper first and hide it.

See About the Assessment in this Unit (p. I-19) and the **Teacher Note,** Keeping Track of Students' Work (p. 11), for further information about observing and assessing students as they work in class.

Activity

Teacher Checkpoint

Representations of Sorted Shapes

Students work individually during this activity. Distribute a large sheet of paper to each student, and ask them to share the Attribute Shape Cards from the previous activity. Have glue sticks available.

You will use this big paper to make a representation of one good way of sorting the shapes. After you have tried a few different ways of grouping the shapes, decide on a favorite way that you'd like to glue down. You might want to circle the shapes that go together, so that other people can tell which belong together. Then write your secret rule on a small piece of paper.

When you finish, you can use these representations to play more Guess My Rule with Shapes.

As students work on creating a Guess My Rule representation, ask them to check in with you before they glue down their shapes. Encourage them to try more than one way of sorting before they glue their cards down. Some students may not yet be sorting by one consistent rule for their group; this is important information for you to note.

When most students are finished gluing their sorted shapes, brainstorm a few possible ways of recording their secret rules. Draw an example of sorted shapes on the board, or draw attention to one student's sorting:

Suppose you had sorted your blocks like this. What are some ways you could write your secret rule?

Students may have difficulty putting their ideas into words. Again, you can help them structure their rule sentence by writing possible frameworks on the board. If writing is too difficult, some students could record their rule visually by gluing down on their large paper *only* the shapes that fit their rule. Or, they could dictate their rule to you or an adult helper.

Distribute small paper scraps for writing the students' sorting rules, and make tape available for taping the rule to their representation.

When you have recorded your rule on one of these little papers, turn it over so that no one can see it. Then tape one edge of this paper to your large sheet. You can also use a paper clip to hold it down. This way, people can try to guess your secret rule on their own, then turn over your rule to see if they were right.

As students finish taping their rules in place, pair them with someone else who has finished to guess their secret rule for sorting. If there are extra Attribute Shape Cards, students may want to use these for guessing the rule, as they have done previously.

When both partners have guessed each other's rules, they find a third student to look at their sorted shapes and guess their rules.

Observing the Students

As students are working, walk around and examine how they have sorted their shapes. Place one or two Attribute Shape Cards in their groups to show them you can guess their rule. Also ask them about their sorting decisions:

How did you sort your shapes? Which shapes go together? How did you decide?

As you are interacting with students, watch for the following:

- Are students identifying a single attribute to sort by? Are they able to apply this rule to all the shapes they are sorting, or do they change the rule midway through the sorting process?
- Are their shapes sorted into clear groups, so that another person can tell which shapes belong in each group?
- How are students guessing each other's rules? Are they able to use the information they get from a right or wrong guess when they make successive guesses?

Sessions 1 and 2 Follow-Up

Family Connection Send home the signed family letter or the *Investigations* at Home booklet to introduce your work in this data unit. If necessary, include a request for lids and buttons to be brought from home for further class work with sorting.

Keeping Track of Students' Work

Teacher Note

Throughout the *Investigations* curriculum, there are numerous opportunities to observe students as they work. Teacher observations are an important part of ongoing assessment. A single observation is like a snapshot of a student's experience with a particular activity, but when considered over time, a collection of these snapshots provides an informative and detailed picture of a student. Such observations can be useful in documenting and assessing student's growth, as well as in planning curriculum. They offer important sources of information when preparing for parent conferences or writing student reports.

The way you observe students will vary throughout the year. At times you may be interested in particular problem-solving strategies that students are developing. Other times, you might want to observe how students use or do not use materials for solving problems. You may want to focus on how students interact when working in pairs or groups. Or you may be interested in noting the strategy that a student uses when playing a game during Choice Time. Class discussions also provide many opportunities to take note of student ideas and thinking.

You will probably need some sort of system to record and keep track of your observations. While a few ideas and suggestions are offered here, it's important to find a record-keeping system that works for you. All too often, keeping observation notes on a class of 28–32 students can quickly become overwhelming and time-consuming.

A class list of names is one convenient way of jotting down your observations. Since the space is somewhat limited, it is not possible to write lengthy notes; however, over time, these short observations provide important information.

Another common approach is to keep a supply of adhesive address labels on clipboards around the room. After taking notes on individual students, you can peel off each label and stick it in the appropriate student's file.

Some teachers keep a loose-leaf notebook with a page for each student. When something about a student's thinking strikes them as important, they jot down brief notes and the date.

You may find that writing notes at the end of each week works well for you. Some teachers find this a useful way of reflecting on individual students, on the curriculum, and on the class as a whole. Planning for the next week's activities often grows out of these weekly reflections.

In addition to your own notes, you will have each student's folder of work for the unit. This documentation of their experiences can help you keep track of your students, assess their growth over time, and communicate this information to others. An activity at the end of each unit, Choosing Student Work to Save, suggests particular pieces of work you might keep in a portfolio of work for the year.

What Did You Notice?

This class has just spent a short time freely exploring a new material, attribute blocks. Now they are sharing some of their observations, which the teacher records on chart paper.

Let's take a couple of minutes to write down what you've been noticing. I heard a lot of observations while I was walking around.

Garrett: Um... Some are flat and some of them can stand up. *[He shows the class what happens when he tries to stand different blocks on edge.]*

OK, so the flat ones can't stand up, but the thick ones can. Other observations, or things you noticed about these blocks?

Jonah: There's a hexagon.

Good for you for remembering that name. Who else noticed some of the other shapes?

Susanna: Triangles. A big one and a small one.

Iris: Squares. Two of them, too.

Nadia: There were shapes of a door.

[The teacher holds up a rectangle.] **Nadia says that there are shapes that look like a door. What a good way of describing that shape, of bringing it into your mind! Does anyone know what this shape is called?**

Donte: Rectangle.

Max: Circle.

Nadia: This shape. *[She holds up a square.]*

Oh, you meant this. *[The teacher holds up a square.]* **What do we call this shape?**

Leah: A square.

How do you know it's a square?

Mia: It has four corners.

Jonah: It looks a little like a book.

So what's the difference between a square *[holds up a square]* **and a rectangle** *[holds up a rectangle]?*

Shavonne: A rectangle's like a door, and longer, and a square's like a piece of paper.

Susanna: If a rectangle was made out of rubber and you squished it down to the size of a square, it would really look like a square, and if you stretched the square out, it would be like a rectangle.

What can you tell me about the sides of the square?

Susanna: On the square all the sides, like how you make the square, are the same, and on the rectangles there's two different kinds.

Can anybody tell me what Susanna is saying?

Jonah: Susanna was saying the lengths on the square are all the same and on the other, on the... *[he forgets the name of the other shape, until someone supplies it]* and on the *rectangle,* there are two different.

Good observing here! What else did you notice about the blocks?

Mia: They have matches, like a big circle and a little circle. And there's a little hexagon and a big hexagon, and a little square and a big square.

Susanna: I noticed they're all primary colors.

Students in this class are describing the attribute blocks by pointing out thickness, shape, size, and color. If they do not yet know the name of a shape, they point to a block or compare it to a familiar object (a door or a book). This process of finding and naming significant features of objects is an important aspect of data analysis throughout this unit.

Guess My Rule with Shapes

This teacher is introducing the Guess My Rule with Shapes game. She supports students as they guess which attribute blocks belong in a group, identify the teacher's secret rules, and establish their own rules for others to guess.

I'm going to make up a rule, but I'm not going to tell my rule. I'm going to put shapes that go together into a set, and you have to figure out my rule by looking at the blocks. *[The teacher starts a SMALL group and a BIG group on and off a sheet of paper.]*

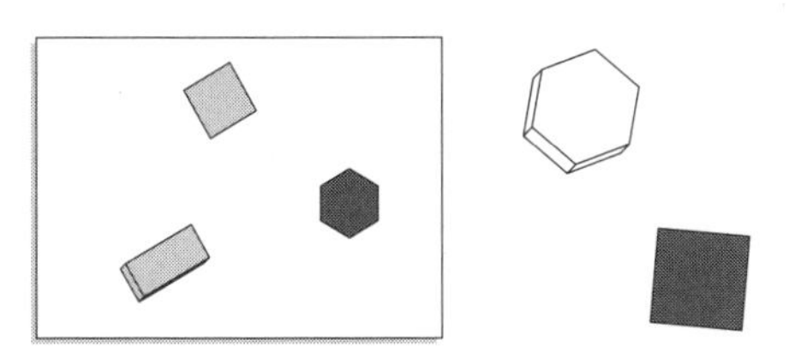

If you think you know my rule, raise your hand, and you can add another block to test your guess. I'll tell you if what you've done fits my rule or if it doesn't.

Andre adds a big thick red triangle to the group off the paper. The teacher says that works with her rule. Max adds a big thick yellow circle to the paper.

That doesn't work with my rule. *[She moves the big thick yellow circle to the BIG group.]*

The next six guesses correctly place other shapes into the two groups. The teacher then asks for a volunteer to say what the rule is.

Andre: Little ones have to go in the little group, and big ones in the big group.

The teacher's next rule is THIN and THICK (or, as students describe it, "a flat group over there and a big group over there." After another round, the teacher then prepares students for making up their own rules.

Now you make up a rule. Look for blocks that would go together in a group. Blocks that go together are *similar* or *alike* in some way. Who sees some that go together?

Chanthou: The big circle and the little circle.

Are there any other blocks here that might go into that group?

Jamaar: The big and little hexagon, because they look kind of the same.

Chanthou *[agreeing]:* Everything that has big and little. Like big triangles and baby triangles.

In Chanthou's set of big-and-littles, we could include every one of these shapes. Does anyone have a group that would include only *some* of these blocks, but not *all* of them?

Chris: The big red rectangle... the little red rectangle... the little red square... *[As Chris points out shapes, the teacher puts them in a group. There are four red shapes in the group, two rectangles and two squares.]*

Chris seems to have a secret rule. Who can guess what his rule might be?

Libby: The rectangle and the square, even though they look different. The square makes the rectangle, but with more on top.

So you could cut a square out of a rectangle, but there's more on top. What else is the same about a square and a rectangle?

Mia: They both have four sides.

Let's count them. Do any other shapes in this set have four sides? *[They check by counting the sides of each other shape.]* **So this could be a set of shapes that have four sides. What else is the same about all of them?**

Brady: Red.

OK. So let's add some more blocks to the group, and Chris will tell us whether or not they also fit his rule. What would be a good block to use to test our ideas?

Brady: A red one.

Why is that a good choice?

Brady: Because we'll know if his rule is red ones.

Session 3

Sorting

Materials

- *Math Counts: Sorting* (optional)
- Class button collection
- Class lid collection
- Other collections for sorting (optional)
- String loops (1 per student, optional)
- Chart paper

What Happens

After reading the book *Math Counts: Sorting* (if available), students discuss sorting objects by different attributes. In a game of Guess My Rule, the teacher sorts the students themselves according to secret rules based on their visible attributes. To prepare for upcoming sessions, students independently explore the class collections of buttons and lids. Their work focuses on:

- sorting people according to one attribute
- using language to describe how a grouping was created (e.g., same/different, particular attributes)
- exploring attributes of buttons and lids
- sorting and categorizing buttons and lids

Activity

Math Counts: Sorting

Use the book *Math Counts: Sorting* to explore sorting with the class. This book presents many color photographs of sets of objects, including buttons, pencils, toys, footwear, tableware, and foods, sometimes sorted in more than one way. As you are reading this book, give students time to answer the questions raised on its pages. Pause to ask:

Why have these objects been grouped like this? How could you sort these out?

In order to respond, students need to figure out how things are alike and describe why they go together. Help students recognize that the same group of objects can be sorted in more than one way, and ask them to identify the author's "rules" for sorting. Write on the board some of the sorting rules they suggest.

If you can't find this book, start this session with the next activity.

Activity

Guess My Rule

We've been playing Guess My Rule with Shapes. One of us grouped the shapes by a secret rule, like YELLOW, and the rest of us tried to guess the rule. Today we are going to play another version of Guess My Rule. Instead of shapes, we'll be grouping the people in our class by a secret rule. I'll show you how it works.

Choose a straightforward, visually obvious rule such as WEARING STRIPES. Choose two students who fit your rule, and when the game starts, ask them to stand in a designated area where the class can see them.

I am thinking of a secret rule. Some people fit my rule and some people don't. To guess what my rule is, you need to look carefully at the people in my group and figure out how they go together.

Chanthou and Garrett, you both fit my rule. Please come and stand together up here. Now, if you think you know what my rule is, don't tell me yet! Just tell me somebody else you think belongs in this group.

Students take turns suggesting someone who might fit the rule. (It is fine for them to name themselves.) If the person named fits your rule, he or she joins the others. Start a second group in another area for people guessed who don't fit your rule.

Point out the importance of all the clues: not just the people who fit the rule, but also those who do not. Be sure students understand that you are grouping people by a characteristic everyone can see, such as hair color or clothing, not a characteristic that can't be seen, such as LIKES CHOCOLATE or HAS A DOG. For more discussion of playing the game with a focus on people, see the **Teacher Note,** Playing Guess My Rule (p. 17).

Keep taking guesses until at least half the students have joined one group or the other. When enough evidence has been gathered and you sense that most students have a good idea of the rule, ask a volunteer to say what he or she thinks the rule is and to give reasons for that guess. If someone finds a rule that fits the evidence but is not the one you had in mind, acknowledge the student's good thinking and explain that finding two rules that both work can accidentally happen sometimes when playing this game.

Play a few more rounds, choosing rules that are visually straightforward. After each rule is guessed, write it on the board in the structure students seem most comfortable with.

Activity

Exploring Lids and Buttons

During the next few days in math class we are going to keep grouping and sorting things. We'll be working with collections of buttons, collections of lids, and *[name any other sorting collections you have].*

For the rest of this session, you may work alone or with a partner to explore these collections. You may sort them however you'd like.

Distribute 10–12 lids or buttons to each student. Explain that the lids and buttons are not to be mixed together. Each child will have a chance to sort both lids and buttons, working first with one type of object, then the other. They may use string loops or sheets of paper to help with sorting.

After a few minutes of exploration, you might stop the class and ask volunteers to describe a few ways they have been sorting the buttons or lids, to give other students ideas. The important thing during this initial exploration is for students to get a sense of what these objects look like.

When students have explored one type of object for about 10 minutes, ask them to exchange with someone who has the other type, so that everyone gets a chance to work with both sets of objects.

At the end of the exploratory time, after materials have been put away, ask students to briefly describe what they noticed about the objects. You might explain that they'll be using these collections in the next few days for more games of Guess My Rule.

Session 3 Follow-Up

Describe a Button Distribute Student Sheet 1, Describe a Button, for homework. Students need to find a loose button from home, a button they are wearing, or one from the class collection (if you have enough) and tape or draw it on this sheet. They then list as many words as they can that describe their button. Family members can be enlisted to help with writing the words. These lists can be hung up around the room or placed in students' math folders after you have looked through them.

❖ **Tip for the Linguistically Diverse Classroom** Students and their families can record the lists of words describing the button in their native language.

Playing Guess My Rule

Teacher Note

Guess My Rule is a classification game in which players try to figure out the common characteristic, or attribute, of a set of people or objects. To play the game, the rule maker (who may be you, a student, or a small group) decides on a secret rule for classifying a particular group of things. For example, rules for people might be WEARING BLUE or HAS BROWN HAIR.

The rule maker starts the game by giving some examples of people who fit the rule—for example, by having two students who are wearing blue stand up. The guessers then try to find other individuals who might fit the rule: "Does Eva fit your rule?"

With each guess, the individual named is added to one group or the other—*does fit* or *does not fit* the rule. Both groups must be clearly visible to the guessers so they can make use of all the evidence—what does and does not fit—as they try to figure out what the rule is.

You'll need to stress two guidelines during play:

- *"Wrong" guesses are clues and can be just as important as "right" guesses.* "No, Brady doesn't fit, but that's important evidence. Think about how he is different from Kristi Ann, Tuan, and Donte." This is a wonderful way to help students learn that errors can be important sources of information.
- *When you think you know what the rule is, test your theory by giving another example, not by revealing the rule.* "Mia, you look like you're sure you know what the rule is. We don't want to give it away yet, so let's test out your theory. Tell me someone who you think fits the rule." Requiring students to add new evidence, rather than making a guess, serves two purposes. It allows students to test their theories without revealing their guess to other students, and it provides more information and more time to think for students who do not yet have a theory.

When students begin choosing rules, they sometimes think of rules either too vague (WEARING DIFFERENT COLORS) or too hard to guess (HAS A PIECE OF THREAD HANGING FROM HIS SHIRT). Guide and support students in choosing rules that are "medium hard"—not so obvious that everyone will see them immediately, but not so hard that no one will be able to figure them out.

Students should be clear about who would fit their rule and who would not fit; this eliminates rules like WEARING DIFFERENT COLORS, which everyone will probably fit. It's also important to pick a rule about something people can observe. One rule for classifying might be LIKES BASEBALL, but no one will be able to guess this rule by just looking.

Guess My Rule can be dramatic. Keep the mystery and drama high with your remarks. "That was an important clue!" "This is very tricky." "I think Diego has a good idea now." "I bet I know what Kaneisha's theory is."

It is surprising how hard it can be to guess what seems to be an obvious rule (like WEARING GREEN). You can't always predict which rules will be difficult. Sometimes a rule you think will be tough is guessed right away; other times, a rule that seemed obvious will turn out to be impossible.

Give additional clues when students are truly stuck. For example, one teacher chose WEARING BUTTONS as the rule. All students had been placed in one of the two groups, but still no one could guess. So the teacher moved among the students, drawing attention to each in turn: "Look carefully at Max's shirt. Now I'm going to turn Claire around to the back, like this—see what you can see. Look along Tamika's arms." Finally, students guessed the rule.

Classification is a process used in many disciplines, and you can easily adapt Guess My Rule to other subject areas. Animals, books, vehicles, tools, and types of food can all be classified in different ways.

Session 4

Sorting Buttons

Materials

- *The Button Box* (optional)
- Set of 20 different buttons
- Class button collection
- Chart paper
- String loops (1 per pair, optional)

What Happens

Students discuss the attributes of buttons, reading together *The Button Box* (if available). They then play Guess My Rule with Buttons. Their work focuses on:

- exploring and describing attributes of object collections
- sorting materials according to one attribute
- identifying like attributes within a set
- categorizing data so that others can guess the sorting rule

Activity

The Button Box

Do any of you have buttons on your clothes today? What does your button look like?

Students describe any buttons they are wearing. You might pair students who don't have buttons with students who do, so that those without can describe their partner's buttons.

We're going to look at a book called *The Button Box.* This book is about a boy who likes to play with and sort the buttons in his grandmother's special box. As we read, notice if any buttons in the book are like the ones that you are wearing today.

This book describes and illustrates buttons according to many categories (painted, sparkly, cloth-covered, leather, pearly, wood, glass, bone, bumpy, smooth), and will give students ideas about attributes to look for in your class button collection.

As you read this book, ask students to describe how certain pictured buttons are alike. If you can't get this book, you might instead take a look in class at the homework students did after Session 3. Talk about the different words students found to describe their buttons, and the categories that those words suggest.

Activity

Guess My Rule with Buttons

You will need the set of 20 different buttons (no exact matches) you have prepared for this session. Seat students where they can see each other and any materials displayed (if possible, in a circle on the floor with buttons in the middle). Choose a secret rule (such as RED buttons) and group together three or four buttons that fit the rule. As before, you might use a sheet of paper or a loop of string to define the groups more clearly. Keep extra buttons nearby for students' guesses.

See the buttons I've just put in this group? They all fit my secret rule. This is like the game we played yesterday, when we sorted people. All the buttons in this group are the same in some way. Can anyone find another button that fits my rule?

Give everyone a moment to look carefully at your buttons. Then call on someone to choose another button that might fit your group. If students need help, you might limit the possibilities by holding up two or three buttons; ask a student to place one of these in the group. If the button guessed does not fit your rule, ask the student to place it in a NOT group, as you did when sorting the students in the previous session, so that everyone can use this information while they are guessing.

Students continue guessing, taking turns, until at least four or five buttons have been correctly placed. Ask someone to guess the rule. All those who agree raise their hands. The student who guessed the rule and anyone who disagrees with that guess should explain their thinking to the class. If everyone wants to guess, students might whisper the rule altogether into the center of the circle.

Clear away the first group of buttons and choose another rule (such as BUTTONS WITH HOLES). Group together three or four buttons that fit your rule.

Now we're going to play again, using a new secret button rule. I've sorted the buttons another way. Take a good look. See if you can guess my rule by choosing a button that belongs in my group.

Continue with more rounds until you have a sense that students are sure about how to play. As you play, remind students that the idea for this game is to sort as clearly as possible, with no tricks, because you are hoping that someone else can guess your rule. This understanding will be important when students play this game in pairs for the next activity.

Activity

Guess My Partner's Rule with Buttons

From the class button collection, distribute 10–12 buttons to each pair of students. You might also provide paper or string to help students define their groups more clearly.

Now you will work with partners to play a guessing game like the one we just played together. For example, Luis might pick a secret rule like METAL, and put three metal buttons in a group. Claire will try putting in some more buttons that she thinks fit his rule. After she places three buttons, she can try to guess the rule. If she does not guess it, Luis puts the button in the NOT group, and Claire keeps trying with other buttons.

Reemphasize that although people may try to trick each other in some guessing games, this game doesn't work that way. You both win only when the other person guesses your rule correctly.

Observing the Students

As you walk around the room, notice how students are sorting their buttons.

- Are they grouping sets of buttons that are alike in some way?
- What strategies are students using to guess someone else's rule?
- Can students choose a clear category and sort their whole set of buttons according to this category? If not, how are they grouping their buttons?

See the **Teacher Note,** Sorting into Groups (p. 22), for more ideas about what to look for as students are sorting.

If students are having difficulty, you might further limit the number of buttons they are working with. Try whispering a sorting rule in their ear and watch how they go about sorting according to this rule. Or model the sorting process by grouping the buttons according to a rule that seems obvious (e.g., RED); this will, of course, vary depending on the available buttons. Then mix the buttons up and ask the student to sort according to another color rule that seems obvious (e.g., BLACK buttons).

Toward the end of the session, ask students to write one of their sorting rules for buttons. Help students structure their rule sentence by writing an example on the board:

My secret rule is ______________

I put together all ______________

Alternatively, students might dictate their rule to you, or they might draw a picture showing only buttons that fit their rule.

Session 4 Follow-Up

Sorting Things Distribute Student Sheet 2, Sorting Things, for homework. Read through the sheet to explain the assignment. Students are to find a group of 10 similar objects at home and sort them into two groups. They should try sorting the same group of objects a few different ways before choosing one way to record on the back of the sheet. To record, they draw the objects in the two groups and write the sorting rule.

❖ **Tip for the Linguistically Diverse Classroom** Students and their families might write the sorting rule in their native language.

Teacher Note — *Sorting into Groups*

As students are sorting their buttons and lids, notice how they create categories. Some students may sort lids by color, putting the white ones in one group and "the rest" in another group. Some may spontaneously sort into more than two groups (e.g., WHITE lids, BLACK lids, BLUE lids, and the rest).

Students often find it easier to guess the teacher's rule than to come up with rules of their own. Many first graders will not use one rule for sorting the whole data set. For example, they may put BIG buttons in one group and RED buttons in another group. Some students may begin by looking for exact matches. As first graders spend more time exploring materials and talking about their features (or attributes), they will become familiar with a variety of sorting rules.

With "real-world" data sets, such as buttons or lids, there are many different attributes to be identified. Further, many of the attributes are "fuzzy"; that is, they are not as easy to define as those of the attribute blocks. There is no disagreement about whether an attribute block is RED or BLUE, but there is much room for discussion about which lids are SMALL or which buttons HAVE HOLES (do the ones on the back of shank buttons count?).

Some of the rules students have invented include these:

- LIDS THAT SOUND LIKE A DOORBELL
- LIDS THAT ROLL
- MY FAVORITE BUTTONS
- DARK COLORED BUTTONS

Rules like these raise legitimate problems of definition and clarification. Some of these issues can stimulate deeper thinking about classification and need not be simplified for students. However, students may need reminding that, just as when sorting people, they must sort in a way that is visually obvious to someone else. Thus, MY FAVORITES is not a rule that someone else can see.

Several particular issues often arise as students of this age work on sorting, all related to the problem of defining "fuzzy" categories.

Sometimes, a category that can be seen is still very difficult to describe. For example, one pair of students chose the rule LIDS YOU CAN ROLL. They demonstrated how to put your finger on the edge of the lid and flip it up so that it could roll on its edge. They had constructed a legitimate "fuzzy" characteristic, hard to define yet real. It was not clear if they meant LIDS THAT COULD ROLL or LIDS THAT COULD FLIP UP.

It is very tempting to "help" students with such an idea by getting them to substitute a simpler rule, one that is not so hard to define. A better approach is to help students clarify their rule simply by asking them which objects do and do not belong in their category, and why.

Some descriptions (like those of size) require a basis for comparison. For example, some students chose LIDS THAT ARE SMALL. When they began to field other students' guesses, they found they could not really agree on which lids did and did not fit their rule. When this happens, the teacher's role is to help students clarify their rule as suggested above. Before they challenge the rest of the class, ask them to decide which objects fit their rule, which objects don't fit, and why. Challenge their choices and help them verbalize their reasons. Even a measurement that seems exact may require further definition.

Another problem for first graders is that different words can be used to describe the same characteristic. The pair whose rule was BUTTONS THAT ARE SHINY did not accept student guesses of REFLECTING BUTTONS, METALLIC BUTTONS, or GOLD AND SILVER BUTTONS. There was outraged protest when they announced their rule because other students felt they had already identified the category in different words.

The teacher helped the student guessers and the student rule-makers explore together whether or not their definitions were really different. In this case, the rule-makers admitted that REFLECTING BUTTONS are shiny, but they presented a good case that GOLD AND SILVER BUTTONS may leave out certain SHINY buttons (such as copper or jeweled buttons).

In such a case, the teacher must encourage the clarification of real differences in definition, but at the same time discourage rule-makers from being too rigid about what they accept as correct guesses. If other students have guessed the concept and have accurately described the characteristics of the category, the exact wording should not be an issue.

Sometimes during sorting activities, students are asked to record their sorting rule. To conceptualize a rule in words may be helpful to some students as they are sorting. But for many others, formulating a verbal rule may be more difficult than grouping the objects. If students have difficulty putting their idea into words, you might give them a simple framework "My secret rule is [blocks] that are ________." For second-language students or any for whom writing is especially difficult, you might model ways to record their rule visually. For example, they glue or record on their paper only the objects that fit their rule.

Figuring out which objects belong together is a compelling and challenging topic for first graders. It is one that they will have many more opportunities to explore through later experiences in both math and science.

Session 5

Sorting Lids

Materials

- Chart paper
- Two lids selected for comparison
- Set of 20 different lids
- Class lid collection
- String loops (1 per pair, optional)

What Happens

As a class, students compare two lids, describing how they are the same and how they are different. This leads to the game Guess My Rule with Lids, with more secret sorting rules. Student work focuses on:

- exploring and describing attributes of object collections
- identifying like features of objects
- categorizing data so that others can guess the sorting rule

Activity

Comparing Lids

Students sit in a circle on the floor or in another arrangement in which they can see each other and the materials displayed.

I have two lids that I will pass around. As you look at these two lids, think about how they are the same. What do they have in common?

Depending on the lids you have chosen, students might see similarities in size, color, material, shape, purpose, the printing on them, and so forth.

Same	Different
have words	1 has picture
have white on them	1 is flatter
round	1 has red
lids that you turn	size

Record the students' ideas on chart paper under a heading of *Same.* If they suggest features that are different, ask them to save that idea for the next part of the discussion, or start a *Different* listing at the same time.

Now look again at these lids and tell me what's different about them. How is one different from the other?

Record students' ideas on the chart under the heading *Different.*

Activity

Guess My Rule with Lids

Gather the prepared set of 20 lids with no exact matches.

So far we've sorted shape blocks, and people, and buttons by secret rules. Now we're going to sort this collection of lids.

Choose a secret rule (such as BIG LIDS) and find three or four lids that fit your rule. Group them in the center of the circle (perhaps on a sheet of paper or in a string loop). Have extra lids that students can use for guessing.

All the lids in this group are the same in some way. Can anyone find another lid that fits my rule?

After everyone has a moment to look carefully at your grouping, call on someone to choose another lid and place it in your group. Or, hold up two or three lids and ask a student to place one of these in the group. If the lid guessed does not fit your rule, ask the student to place it in a NOT group.

When students have correctly placed at least four or five lids, ask someone to guess the rule.

What information did you use to figure out the rule? What do the rest of you think? Does anyone disagree? Why? Are you convinced this is the right rule?

Play again using a new rule, such as CLEAR LIDS or LIDS WITH WORDS. Continue playing until you feel that students will be able to play the game as partners. The **Dialogue Box,** Sorting Lids (p. 28), demonstrates how one teacher modified this introductory activity to help students focus on the attribute chosen for the secret rule.

Activity

Guess My Partner's Rule with Lids

Give each pair of students about 15 lids and (optionally) paper or string to help define their groups.

Now you're going to play this same game with a partner. Let's say Yukiko decides on the secret rule PLASTIC LIDS. She groups three or four plastic lids together. Tony tries to find more lids that fit her rule. After he places three lids, either in Yukiko's group or in the NOT group, he can guess the rule using words. If he guesses the wrong rule, he keeps placing lids until he figures it out.

Remind students that in this game, they are not trying to trick their partner; they need to sort clearly because they both win when the rule is correctly guessed.

Observing the Students

As you walk around the room, notice which attributes students are using to sort their lids.

- Are students able to group sets of lids that are alike in some way?
- What strategies are students using to guess their partner's rule?
- Can students choose a clear category and sort their whole set of lids according to this category? If not, how do they group the lids?

If time permits, partners might work together to sort lids according to another secret rule, then ask another pair of students to guess their rule.

Toward the end of the session, ask students to record one of their sorting rules for lids. For example:

My secret rule is lids that ______________

Alternatively, students might dictate their rule to you, or draw a picture of only those lids that fit their rule.

Session 5 Follow-Up

Homework

Design a Lid On Student Sheet 3, students design their own lid, drawing it extra-large on a sheet of paper. If they want, they can copy from a real lid they find at home. They then list some of the different groups their lid would fit in, if it were added to the class collection. For example, a student's list might include groups like PLASTIC LIDS, LIDS WITH WRITING, and LIDS THAT SNAP ON.

❖ **Tip for the Linguistically Diverse Classroom** Students and their families can do the written work in their native language.

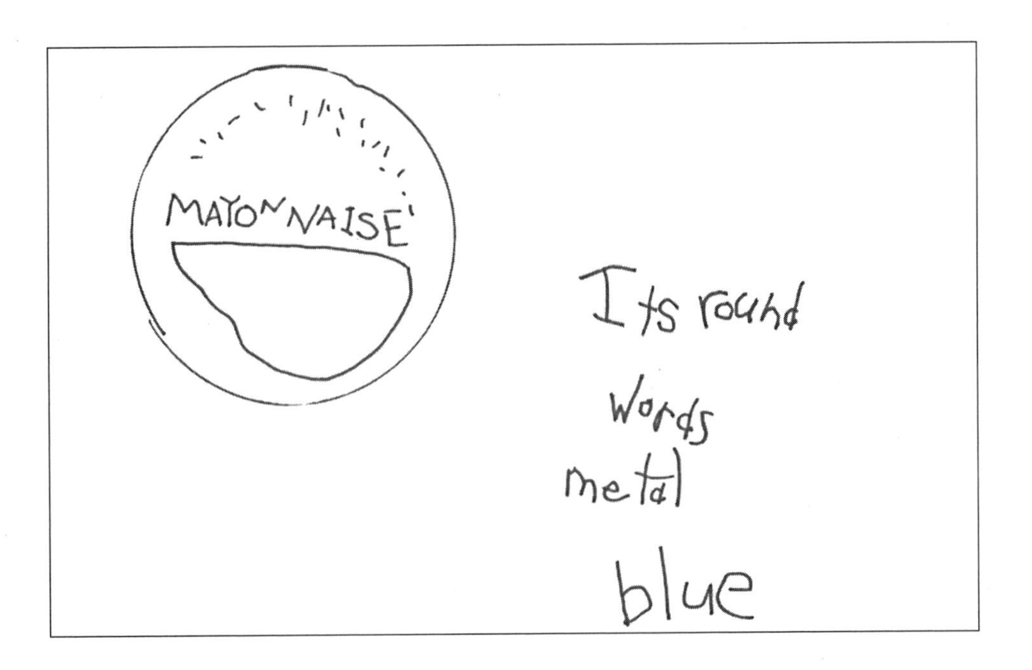

Extension

Same and Different Working in pairs or small groups, students choose two lids to compare and tape them to a sheet of paper. Below them, they write their ideas in two lists headed with the words *Same* and *Different.* If writing is too difficult and you have classroom help, students might dictate their ideas. Display their finished comparisons around the room.

DIALOGUE BOX

Sorting Lids

As this teacher introduces Guess My Rule with Lids, she varies the game by letting students guess the rule in words from the very beginning. Once they guess it, she asks students to continue finding lids that fit her rule (in this case, LIDS WITH WORDS). By choosing another lid that fits the rule from among the extra lids, students practice focusing on one attribute at a time while ignoring the other attributes.

I'm going to sort some of these lids into a group. See if you can tell why I put those lids together. Try to guess what my secret rule is.

The teacher puts seven lids on a sheet of paper, all facing top up: large and small metal juice lids, large and small yogurt cup lids, a white plastic lid with writing, a spice jar lid, and a milk lid.

All right. These are the lids I chose to go on my paper for a specific reason. There's something that's the same about all of them. What do you think? What's my secret rule?

Tuan: Matches? *[He is looking at the pairs of juice and yogurt lids.]*

Hm, there are some that match, but all seven lids have something the same. There's one reason I chose them all for this group.

Nadia: They all have words.

[To check Nadia's guess, the teacher holds up each lid in turn; students agree that all the lids do have words on them.] **You guessed my secret rule: LIDS THAT HAVE WORDS. Libby, choose another lid that should go in my group.**

[Libby stares for a long moment at the lids on the table.]

What are you looking for?

Libby: *[Pause]*...Writing.

Finally Libby picks one and adds it to the group. A few other students choose lids with writing to add to the group. When one student picks a lid with just a bar code (no words) and places it on the paper, the teacher moves it to the NOT group.

These are the lids that do NOT fit my rule. Why doesn't this lid fit?

Tony: Because there are lines, but no words.

We have a lot of lids here now, all LIDS WITH WORDS. That was my first secret rule. But there are lots of different secret rules we can use for lids. I'll dump these off my paper, and then we'll start again with a brand new rule.

Representations of Sorted Objects

What Happens

In an assessment activity, students sort buttons or lids and create a representation of their sorted groups. The sorting rule is recorded on a card that is fastened, facedown, to the representation, so that this student work can be used for further games of Guess My Rule. Their work focuses on:

- identifying and describing attributes of lids and buttons
- sorting objects according to one attribute
- creating a representation and a rule for a set of sorted objects

Materials

- Class lid and button collections
- Large paper (1 sheet per student)
- Glue sticks, clear tape (to share)
- Paper scraps, about 3 by 5 inches (1 per student)
- Paper clips (1 per student)

Activity

Assessment

A Favorite Sorting Rule

Students may work at tables, desks, or any place where they will have room to spread out their materials.

Earlier, each of you made a big picture, or *representation,* of one way to sort the shape blocks. Today you'll be doing the same thing with either lids or buttons. I'll give you large paper, and you'll make a representation of one of your favorite ways of sorting.

Take some time before you choose a way to sort. Suppose you decided to sort buttons. You might try putting all the buttons with NO HOLES in one group on your paper, and the ones with HOLES in another group. That's one idea—but then you might try another way, like sorting by colors. Keep sorting until you find a way you really like.

What if you wanted to sort lids? What are some ways you could do that?

As students suggest ways of sorting, invite them to organize a few lids to reflect their idea. After other students check to see if the rule is being followed, they may suggest other ideas.

Once you've picked your favorite sorting rule, you'll need to make a representation. If you are using buttons, you will draw each button to show where it was on your paper and how it looks. If you use the lids, you may glue down your actual lids. Just be certain of your rule before you start gluing.

When you have finished your representation, write your sorting rule on a small piece of paper *[hold up a sample].* **Tape one edge of this paper to your large sheet, writing side down, so it can be a secret rule for people to guess.**

Students work independently. Each needs 12–15 lids or buttons, a large sheet of paper, and a 3-by-6-inch scrap of paper for the rule. Students working with buttons will need pencils or markers for drawing; those working with lids will need glue sticks. You might ask students to check in with you before they begin to draw their buttons or glue down their lids, to make sure that they tried different ways of sorting before choosing one.

Note: Some students may express frustration over the difficulty of drawing buttons, especially if they are trying to show such attributes as thickness, shininess, or specific colors. These students might be happier working with the lids.

As students begin to finish their representations, you may need to remind them about how to record their rule and how to attach it to their large sheet. You might leave rule frameworks on the board for reference:

My secret rule is ______________

I put together all ______________

Distribute paper clips and tape as students are ready for them.

❖ **Tip for the Linguistically Diverse Classroom** Offer support as needed with the recording of rules.

When you have written your rule on a small paper, turn it facedown so we can't see it. Then tape one edge of it to your large sheet. You can also use a paper clip to hold it down. This way people can guess your secret rule on their own before they read what you wrote.

Students who finish early might work with a different set of objects and create a second representation. If there isn't enough time for that, suggest that they look at other students' representations and try to guess the secret rules.

As you look at students' representations, compare them to the representations of sorted shapes that they made earlier in this investigation. Consider the following:

- Can students choose a clear category and sort an entire set of objects according to their rule? Do they use a NOT group?
- Are students' representations clear to others? Are their groups distinctive?
- Do students' written rules reflect the grouping that their representation shows?

Display these representations where students can later play Guess My Rule with them. Keep in mind that button representations may be harder for students to guess, because certain attributes are difficult to record.

INVESTIGATION 2

Survey Questions

What Happens

Sessions 1 and 2: Would You Rather? Students respond to a survey question that the teacher poses, collect data (answers) from the whole class, and figure out ways to represent that data with cubes, drawings, or other materials. Each pair then gets another survey question to investigate, and they work as partners to figure out a plan for gathering their own data.

Sessions 3 and 4: More Surveys and Sorting In a Choice Time format, students work on three activities: two that involve sorting, and one for collecting and recording data that they will use in the next two sessions.

Sessions 5 and 6: Representing and Sharing Survey Results Using the data they collected during Choice Time, students create a representation of the results of their surveys. Each pair then makes an oral presentation of their findings to the whole group.

Routines Refer to the section About Classroom Routines (pp. 100–107) for suggestions on integrating into the school day regular practice of mathematical skills in counting, exploring data, and understanding time and changes.

Mathematical Emphasis

- Making a plan for gathering and recording data
- Sorting and categorizing data
- Inventing and constructing representations of data
- Explaining and interpreting results of surveys
- Presenting data to others in a way that communicates information clearly
- Making sense of other students' representations

You'd rather eat spaghetti than pizza. That's another S.

What to Plan Ahead of Time

Materials

- Interlocking cubes: 500 (Sessions 1–2)
- Stick-on notes (Sessions 1–2)
- Survey boards and Kid Pins (Sessions 1–2, optional)
- Button and lid collections: about 50 of each (Sessions 3–4)
- Attribute blocks and any Attribute Shape Cards left from Investigation 1 (Sessions 3–4)
- String, cut into two-foot lengths and tied in loops (Sessions 3–4, optional)
- Large paper (11 by 17 inches): 1 sheet per student, plus extras (Sessions 5–6)
- Dot stickers, squares of colored paper (Sessions 5–6, optional)
- Markers, glue (Sessions 5–6)

Other Preparation

- If you do not have Kid Pins and survey boards from *Mathematical Thinking at Grade 1,* you can use cubes for gathering data. Prepare sets of cubes in two colors, with enough for each student. Have extras available in a third color. (Sessions 1–2)
- Find pictures of eagles and whales to help students think about the survey question. (Sessions 1–2, optional)
- Prepare a one-column class list, 2 copies per student, plus extras. (Sessions 1–4)
- Prepare three Not-Boxes for Choice Time by stapling or taping together two flat, open boxes or their tops. Label one box in each pair with the word *not.* (Sessions 3–4)
- Prepare several collections of 10–15 items (lids, buttons, or other small objects) in a plastic bag or cup, making 1–2 sets for each Not-Box. Sort one set (e.g., SHINY and NOT SHINY lids) to demonstrate a Not-Box. (Sessions 3–4)
- Make clipboards for student pairs by attaching a large binder clip to a piece of heavy cardboard. (Sessions 3–4, optional)
- The software Sorting with Tabletop, Jr. (see p. 47) offers more sorting experiences. If you have a computer in your classroom, you might make this another option during Choice Time. (Sessions 3–4)
- Duplicate the following student sheets and teaching resources, located at the end of this unit. If you have Student Activity Booklets, copy only the items marked with an asterisk.

For Sessions 1 and 2
Survey Questions* (pp. 118–119): enough to provide 1 question per pair, cut apart. (Choose questions you think will work best with your class.)

Student Sheet 4, Our Plan for Collecting Data (p. 115): 1 per pair

Student Sheet 5, Collections for Sorting (p. 116): 1 per student, homework

For Sessions 3 and 4
Shape Rule Cards* (p. 120): 3 sets, each cut into six individual cards

For Sessions 5 and 6
Student Sheet 6, Our Findings (p. 117): 1 per pair

Sessions 1 and 2

Would You Rather?

Materials

- Eagle and whale pictures (if available)
- Sets of cubes in two colors, or Kid Pins and survey boards
- Class lists (1 per student)
- Stick-on notes
- Survey questions
- Student Sheet 4 (1 per pair)
- Student Sheet 5 (1 per student, homework)

What Happens

Students respond to a survey question that the teacher poses, collect data (answers) from the whole class, and figure out ways to represent that data with cubes, drawings, or other materials. Each pair then gets another survey question to investigate, and they work as partners to figure out a plan for gathering their own data. Student work focuses on:

- inventing representations of data to communicate findings
- explaining and interpreting results of data surveys
- making a plan for gathering data and recording it

Activity

Would You Rather Be an Eagle or a Whale?

We've been sorting objects we can move around, like shapes, and lids, and buttons. Once, we sorted people by what they were wearing or how they looked. Now we're going to try a different way of sorting people—we're going to sort them by what they think or what they do.

We can't usually tell what people think just by looking at them. We'll have to ask them, by taking surveys. Has anyone ever been part of a survey before? What do we use surveys for?

If you and your class have worked with surveys before, remind them what the topic was. For example, in the *Investigations* unit *Mathematical Thinking at Grade 1,* class survey topics included shoelaces (wearing them or not?) and ways to get to school. Spend a few minutes as students share their ideas about how surveys work and what they are used for.

Explain that there are different kinds of surveys. Many focus on serious questions, but others ask more playful questions that give us an interesting glimpse of other people, showing us how people are alike and different. All surveys are ways of gathering data—finding information about, or the opinions of, a group of people. Students' work in this investigation will involve questions they can have some fun with.

In our surveys this week, we'll ask everyone in the class the same what-do-you-think questions. When we have answers from everyone, that will be our data. Then we'll figure out how to sort and show our data, or what we have learned. That way, we can tell how our whole class is thinking about the questions.

Here's our first survey question. Think about this quietly for a moment, and don't tell us your answer until it's your turn: Suppose you could be an eagle or a whale for one day. Which one would you rather be?

If you have pictures of eagles and whales, show them, or briefly brainstorm about the two animals. Ask students to close their eyes, visualize an eagle, and pretend to be that eagle. Do the same with a whale. Keep encouraging students to keep their choices to themselves until it's their turn to provide data.

Note: If you are using Kid Pins and survey boards, label the boards *Eagles* and *Whales,* and follow your customary procedure for gathering student responses. Otherwise, continue the activity using cubes to gather data, as described below.

How could we use cubes to keep track of our survey data?

Students are likely to suggest using one color to stand for eagles and the other color for whales. They may even suggest using particular colors that they associate with each animal. If students insist on a category for people who cannot decide or want to be both, ask them how this data should be represented.

Each of you may come up and take a cube that stands for the animal you would rather be. Then let's put our cubes into towers so that we can tell more easily which group is bigger.

Collect all the cubes of one color from students, creating a tower as you collect each cube. Place these towers on the chalkboard ledge or in the middle of the circle and label them with stick-on notes with words and/or pictures for each category.

Let's talk about the reasons that people prefer to be eagles or whales for a day. Why did some of you want to be eagles? Why did you want to be whales?

Students often want to be whales because they like to dive or swim. Many who prefer to be eagles want to know what it feels like to fly, especially over cliffs.

What have we found out from this survey? Altogether, how many people voted? How do you know? Did everyone vote?

Which is more, the number of people who would rather be eagles, or the number who would rather be whales? How can you tell? Is one a lot more, or are the numbers close?

See the **Dialogue Box,** Discussing Eagles and Whales (p. 42), for the way this discussion went in one first grade class.

Activity

Representations of Eagle and Whale Data

In this activity, students make their own representations of the results of the eagles and whales survey. They might use cubes, stick-on notes, drawings, or any other materials you propose. For information on the types of representations you might expect, see the **Teacher Note,** Inventing Representations (p. 40).

Distribute a class list to each student for making individual records of the data collected in class. Having this at hand will make it easier for them to do their representation. One quick way to record the data is to call out each name on the list; as students register their vote, everyone writes E or W next to each name.

We have lots of materials you could use to create your own representation of the eagle and whale data. If you decide to use paper and pencil, what might you draw?

Students may suggest drawing eagles and whales, drawing the Kid Pins (if you used them), or drawing the faces or names of students who belong in each group.

Could you use cubes? What could you do with cubes to show our eagle and whale data?

One possibility is using one color for eagles and another for whales, keeping the two groups separate. Another idea is to simply use the cubes as counters without regard to color, making a separate stack for each category (eagles and whales) and labeling them clearly.

Could you use stick-on notes? How could you use them to show the same data?

If you have stick-on notes in more than one color, students may suggest using one color for eagles and another for whales. Others may suggest drawing the two animals on individual stick-on notes and grouping them on a larger sheet. Some students may even decide to include the names of their classmates on stick-on notes in their representation.

You may use any of these materials we've talked about to find a way of showing our survey results. Be sure to label your data so that someone else looking at it could tell what we found out.

Observing the Students

As students create representations with materials of their choosing, observe them at work.

- Are students making representations that communicate the results of this survey clearly to others?
- Are they accurate in their counts of the numbers of eagles and whales?
- Are students grouping the data that belong together? Are their categories distinct?

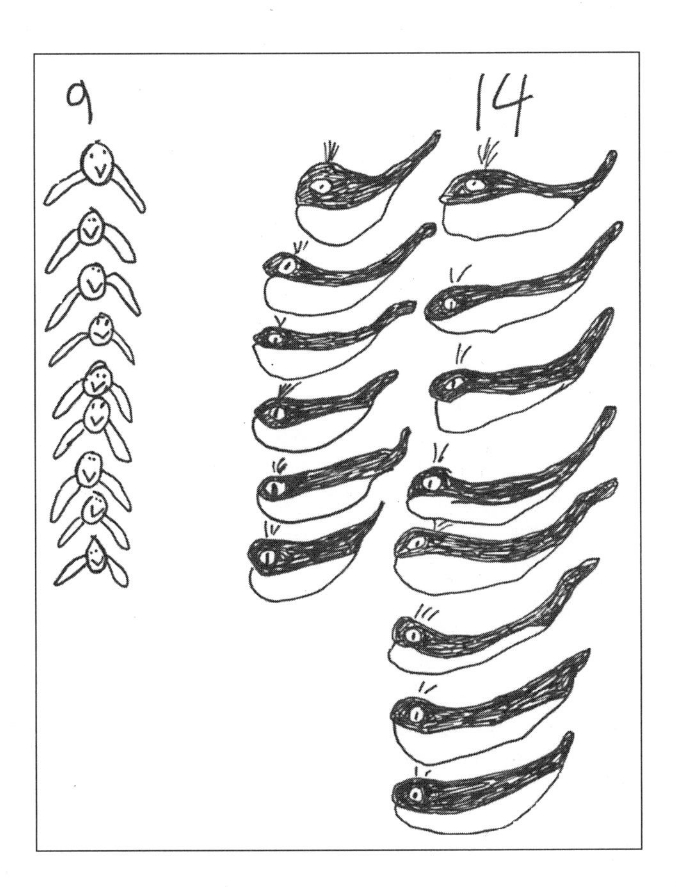

Interpreting Representations When a few students have finished, demonstrate the next step. Two students who have finished trade their work. Each one takes a turn explaining the other person's representation and what it shows.

What will you do if you don't know what something means?

Students can ask the author to explain it, and possibly change the representation to make it clearer for someone else. When students finish working with one partner, they can pair up with someone else and repeat the process as time allows.

Activity

Choosing a Question and Making a Plan

I have some more survey questions. Not everyone will get the same question. You will be working with partners over the next couple of days to collect data from everyone in our class about the question I give you.

Distribute the prepared set of survey questions you have decided to use. You might distribute the questions randomly to partners, or let student pairs draw two cards and choose one. Students might trade questions with another pair if they get a question they don't like.

These questions will be difficult for students to read independently. Encourage them to help each other, and to ask you or another adult for reading help as needed. Some may want to draw a related picture on their card, as a reminder.

❖ **Tip for the Linguistically Diverse Classroom** If you have help from bilingual aides or parents, you might provide translations of the questions (or key words in them). Otherwise, plan to provide pictures and have other students act out parts of the questions to ensure comprehension.

Note: If your students are interested in creating their own questions, be sure to allow plenty of extra time for this step. The process can be difficult, as they need a question that has exactly two possible answers. Working with the whole group, you might list some suggestions on the board to help students get started, then brainstorm other questions. Students may then choose which question they'd like to work on.

When every pair has a question, they are ready for a class list and Student Sheet 4, Our Plan for Collecting Data. Go over the sheet, explaining to students what each question is asking. Take some student ideas for the last two questions: how to record the answers, and how to be sure they have surveyed everyone in the class.

Each team needs to make a plan for collecting data on their question. Decide who is going to ask the questions, who will record the answers, and how you'll record the answers. What are some ways you might do this?

Possible approaches are using the class list to record their data, or asking those surveyed to choose a cube or stick-on note of a certain color to reflect their choice. To be sure they collect data from everyone, students might suggest checking off the names on the class list, or moving through the room in a particular order. This will be a difficult problem for some pairs. Although their initial ideas may not be workable, allow students to discover this for themselves and modify their plan later as the need arises.

Give students the remaining class time to work on their plan and complete Student Sheet 4. For the first question, students might staple their survey question to this sheet, rather than copying it out. If writing out their ideas is too difficult for some pairs, you or another adult might record for them.

Ask students to check in with you when they have finished their plan, to make sure that they are ready to collect data. Pairs who don't complete their plan during this session can use Choice Time during the next session to finish up.

Help students find a place to save their plan, their class list, and anything else they need for the actual collection of survey data. They might use their math folder or a labeled container to keep the materials together.

Sessions 1 and 2 Follow-Up

Collections for Sorting Explain to students that they will be doing some more sorting activities in their next math class. Student Sheet 5, Collections for Sorting, asks them to bring in a collection of 10–15 small items from home that they can use as a sorting material during Choice Time.

More Eagles and Whales Students write a story and draw a picture about their preference to be an eagle or a whale. After collecting the pictures, you might ask students to sort them for display or to put together as two separate books, "I'd Rather Be an Eagle" and "I'd Rather Be a Whale."

Teacher Note

Inventing Representations

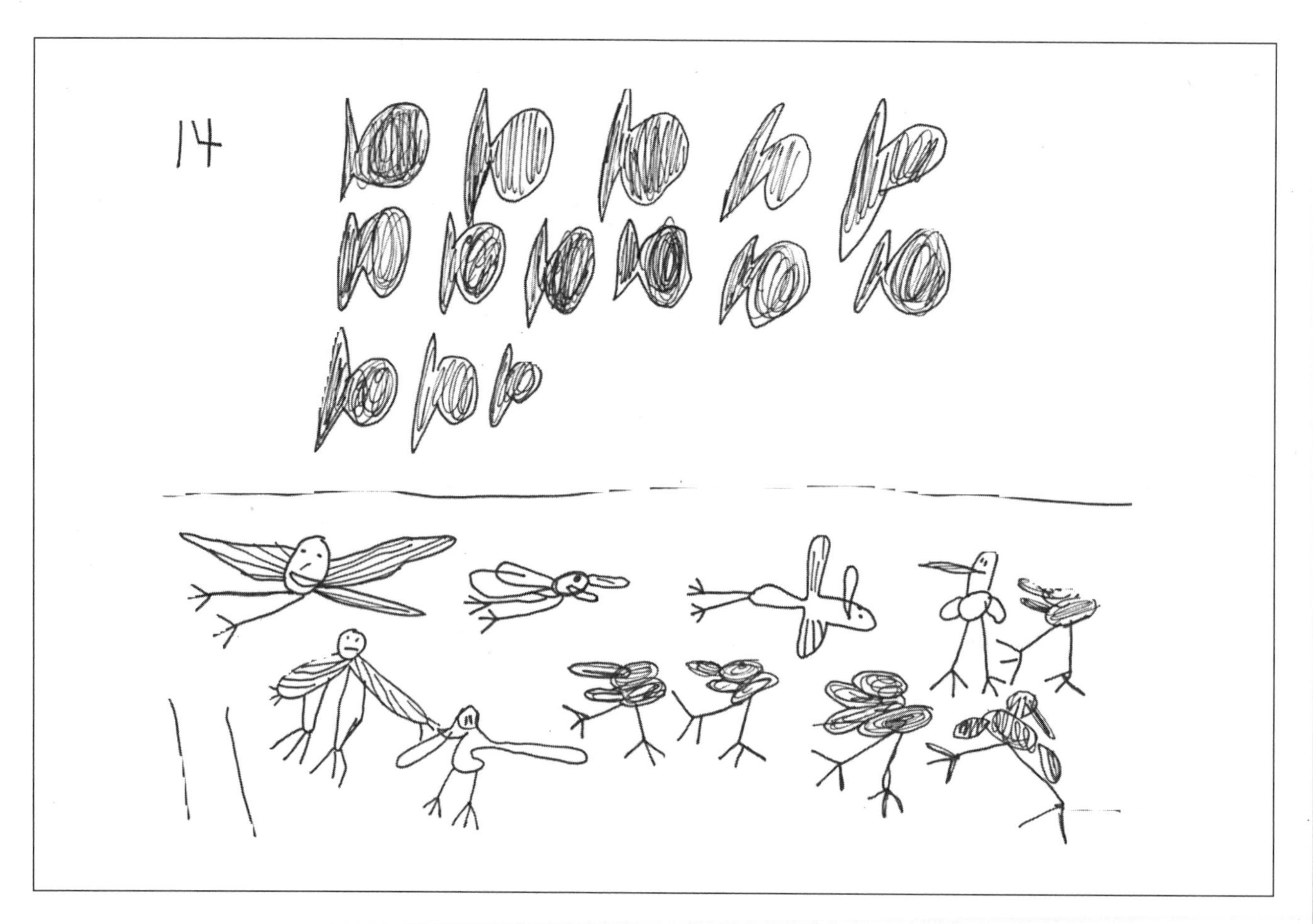

What is a representation? It's a form of communication that students need to learn as part of data analysis; put simply, in the words of one student, "It shows the data." There are "standard" forms of data representation: charts, tallies, line plots, bar graphs. There are also many unusual forms of graph and diagrams; some of the most striking were devised by a statistician or scientist to represent a single, unusual data set in a new way. Even now, new "standard" forms are taking their place in the statistician's repertoire beside the more familiar bar graph.

So, how do students "show the data"? For the activities in this unit, we encourage students to invent their own methods. They start in Investigation 1 by representing objects they have sorted. These first representations may be concrete objects glued in place, or straightforward drawings of objects the students see before them. In the later investigations, when students show class survey data, you may begin to see representations that are partly picture, partly graph.

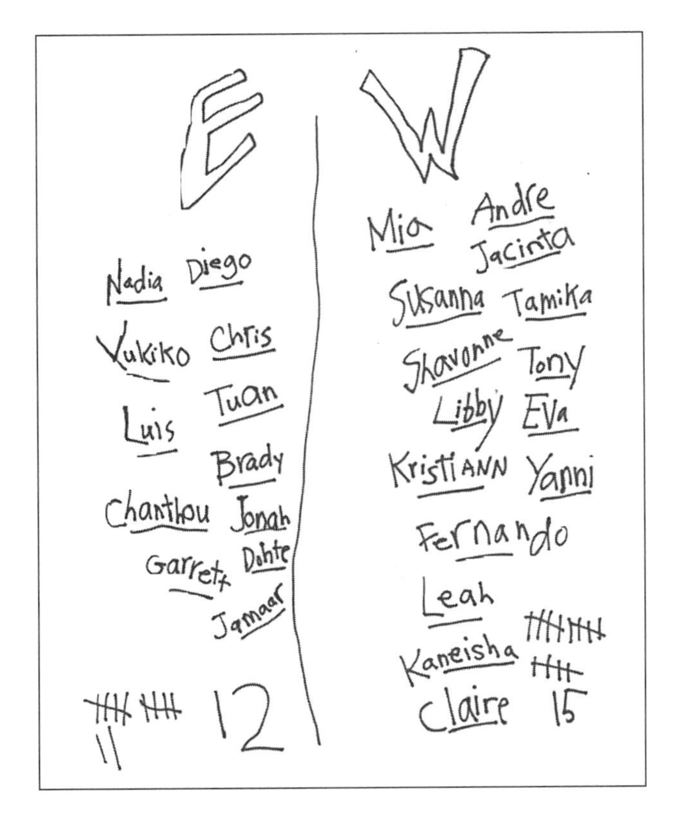

When students invent their own ways of representing their data, both now and in later grades, they often come up with wonderfully individual pictures or graphs that powerfully communicate the meaning of the data. Students' picture-graphs will not necessarily follow the conventions of graphing that adults would use. For example, their pictures may not all be the same size or neatly lined up so that it is easy to compare the relative sizes of different groups. Even so, they may tell a vivid story.

In the *Birthdays* investigation, students are introduced to a simple form of a bar graph for comparing the number of students born in each month. In this graph, each cell is identified by a student's name, and this helps students link the symbolic graph to the data it shows. However, when students are pushed to consistently adopt the conventions of what graphs are "supposed" to look like, they often produce mundane bar graphs which, for them, no longer communicate anything about their data.

Over the years, students will learn to use and interpret commonly used graphs, but for the representations in this unit, picture-graphs have more power. By using their intuition and expectations to construct a meaningful representation, students come to "own" the data. Through this process, they become more familiar with numerical relationships as well as with the connection between the action of counting, the objects they count, and the symbols that represent these quantities.

At first, creating representations will be challenging, but students' skill in organizing and representing data develops with experience. Shown on these pages and on page 37 are some of the ways that first graders have represented their findings from Would You Rather Be an Eagle or a Whale? None of these follow a standard graph form, but all show the data clearly and effectively.

DIALOGUE BOX

Discussing Eagles and Whales

Using Kid Pins and survey boards, this class collected their responses to the question, Would you rather be an eagle or a whale for one day? During this follow-up discussion, the teacher stresses the relationship between the *total* number of children who participated in the survey and the number who answered in each category.

What did we find out from this survey?

Donte: That Jonah and me both wanted to be eagles.

Eva: I thought Diego would say whale, but he didn't.

Nathan: It looks like whales won.

What do you mean, whales won?

Nathan: More kids wanted to be whales.

Tamika: There were 15 whales.

Eva: Yeah, but a lot wanted to be eagles, too.

How many wanted to be eagles?

Eva: It's sort of hard to count them [the Kid Pins] because they're all over the place.

See if you can arrange them so it's easier to count.

[Eva goes to the survey board and puts the eagle Kid Pins into a straighter line. When asked to count them, Eva counts one by one. Another student double-checks.]

Eva: So there's 12 who wanted to be eagles.

Kristi Ann: And Max and Iris couldn't decide what they wanted to be.

Yes, we have two who are undecided. So 15 who wanted to be whales, and 12 who wanted to be eagles. How many is that altogether?

Mia: I think it's going to be a lot, like maybe 50.

Tamika: It should be 30, because that's how many are in our class.

Fernando: But Susanna is absent!

Donte: We could count them. Just count every Kid Pin up there, and then we'll know for sure.

That's one way to see. How else could we find out?

Kristi Ann: Use the number line.

How would that work?

Kristi Ann starts at 16 and counts up 12. Someone else reminds her about the two undecided students. The group concludes that 29 students participated in the survey. They check, using Donte's method, and decide that this makes sense because if Susanna were in class, they'd have their usual class total of 30.

It's important for students to recognize that the number of "eagles" and the number of "whales" (plus those in any other categories students may have created) adds up to the total number of students who voted. Another element of this discussion is a *comparison* of the numbers in each category. Students often use the terminology "winning" to describe which group is bigger. Emphasizing a student's observation of relative size rather than the competitive notion of winning, as this teacher did, keeps the conversation focused on the mathematics.

More Surveys and Sorting

What Happens

In a Choice Time format, students work on three activities: two that involve sorting, and one for collecting and recording data that they will use in the next two sessions. Their work focuses on:

- collecting and keeping track of survey data
- sorting object collections
- guessing how someone else sorted a collection

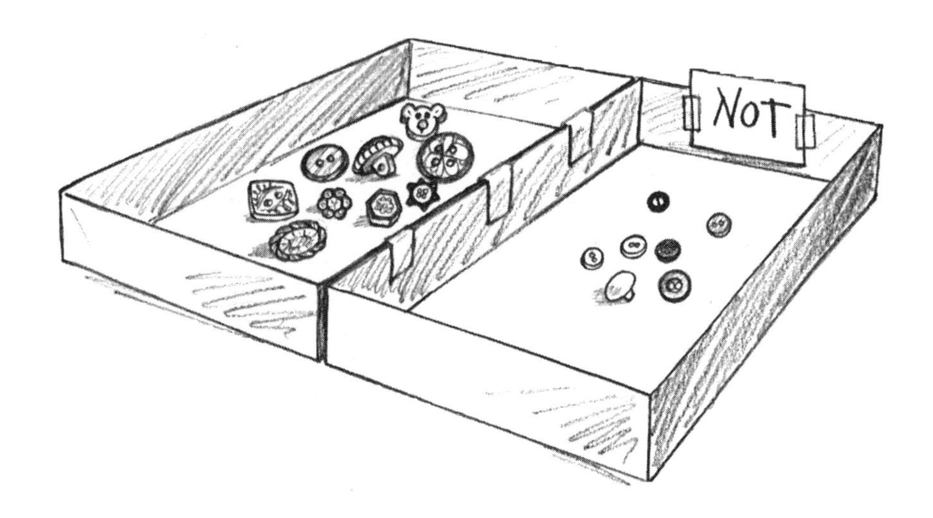

Materials

- Completed plans for collecting data (Student Sheet 4)
- Class lists (1 per pair)
- Cubes, stick-on notes, or dot stickers
- Clipboards (1 per pair, optional)
- Button and lid collections
- Students' sorting collections
- Attribute blocks and Attribute Shape Cards
- Shape Rule Cards (3 sets)
- String loops (optional)
- Not-Boxes

Activity

Not-Boxes

Introduce this new activity by gathering students together and showing them the open, flat boxes that you have used to sort something (such as a small set of buttons, FANCY and NOT FANCY).

We call this a Not-Box. We can use it for sorting and for playing Guess My Rule games. Look at the way I sorted these buttons and see if you can figure out my sorting rule. All the buttons on this side have something in common. All the ones on the other side do *not* have that feature.

Call on a few students to add a button to either box. After a few buttons have been placed correctly and most students seem to know what the rule is, ask someone to describe it.

If students seem to need more experiences to understand the Not-Box, sort the buttons another way (for example, TWO HOLES and NOT TWO HOLES) and play Guess My Rule with Buttons again.

Activity

Introducing Choice Time

Choice Time is a format that recurs throughout the *Investigations* curriculum. During each day of this Choice Time, students get to choose from several activities. They can select the same activity more than once, but they also have to try more than one activity—they can't do the same one over and over, every day.

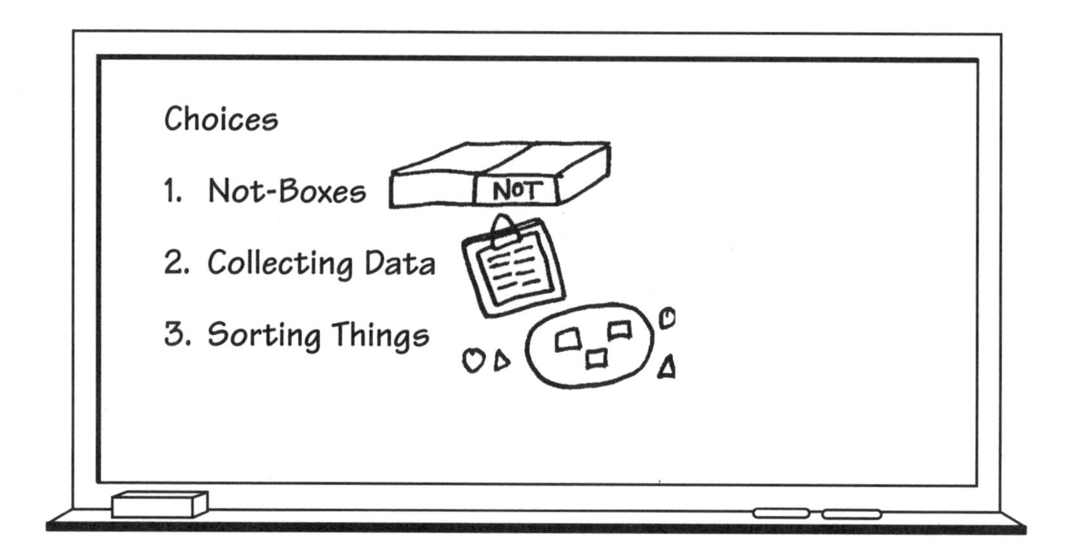

List the activity choices on the board or on a piece of chart paper, with a simple sketch as a nonverbal reminder. You may want to set up the Not-Boxes (Choice 1) and Sorting Things (Choice 3) at different classroom centers, with all the materials students will need. For Collecting Data (Choice 2), partners will be working at their seats or moving around the room surveying classmates.

Today and tomorrow in math class, you'll have Choice Time. There are three activities to choose from. We just talked about Not-Boxes; that's your first choice. You can do the sorting part of Not-Boxes by yourself, but you'll need your partner if you want to play Guess My Rule.

Collecting Data is the second choice. This means taking your surveys. You'll need your plan from yesterday (Student Sheet 4) and a class list. Remember to stay with your partner. Everyone needs to do this choice by the end of our next math class. *[You might put a star by Choice 2 as a reminder that they must do it.]*

The third choice is called Sorting Things. Did anyone bring in a collection from home to use? Do you want other students to use your collection? *[Let students decide whether or not they will share their collections with others.]*

For this third choice, you can also sort shape blocks or the Attribute Shape Cards. There are some Shape Rules *[show these]* **to give you ideas for different rules to sort by. You can also play Guess My Rule with this activity, no matter which materials you use for sorting.**

If you have already established a procedure to help students keep track of the choices they have completed, remind them what they need to do. One possibility is that when they complete an activity, they record its name (or picture) on a sheet of paper which they keep in their math folder. Or, you might post a sheet of lined paper for each choice, headed with the name of the activity and corresponding picture. When students have completed an activity, they print their name on the appropriate sheet.

For the remainder of this session and all of the next session, students work on the three choices. Remind students that while they are working on Choice 1 or Choice 3, they should be available to answer survey questions for classmates who are working on Choice 2.

Choice 1: Not-Boxes

Materials: Prepared Not-Boxes, small collections of lids or buttons (10–15 with each box)

Students use familiar materials and sort them into two groups (for example, RED and NOT RED). After sorting, they play Guess My Rule with a partner. You could post chart paper next to this station for students to record one of their sorting rules.

Choice 2: Collecting Data

Materials: Students' completed copies of Student Sheet 4, Our Plan for Collecting Data; class lists; materials for collecting data (such as cubes, stick-on notes, dot stickers); clipboards (1 per pair, optional)

Generally three or four pairs can work on this choice at the same time. The students' first job is to finish Student Sheet 4, Our Plan for Collecting Data, if they have not already done so. You need to review each plan before they begin their surveys. Student pairs go to each classmate to ask their survey question. When they are sure they have collected and recorded data from everyone in the class, they are finished with this choice. Be sure students save their data in a safe place, as they will work with it further in Sessions 5 and 6.

Note: Some teachers prefer to have all students, or sometimes half the class, collect their data at the same time. If interest in the survey is high, students might begin work on representations of their data as an additional Choice Time activity during Session 4. See Sessions 5 and 6, Making Representations of Our Data, for things to consider during this activity.

Choice 3: Sorting Things

Materials: Attribute blocks and Attribute Shape Cards; students' own collections (optional); Shape Rule Cards; string loops (optional).

Students may sort any way they choose, using two or more categories. String loops may help them organize and define their groups. Students may try sorting different materials, but should not mix up the collections. After sorting, they might find a partner to play Guess My Rule.

Students who sort Attribute Shape Cards or attribute blocks have the option of using the Shape Rule Cards. They pick a card and sort their shapes according to the given rule (for example, by color: red, blue, yellow). As with the other materials, a partner may try guessing the rule by placing more blocks into the groups. Encourage them to place four or five blocks correctly before saying their guess aloud.

Observing the Students

As you observe students during Choice Time, take note of the following:

Not-Boxes and Sorting Things

- Are students identifying clear attributes to sort by?
- How are students grouping objects? Are they able to choose a rule and sort an entire set of objects according to this rule?
- What information are students using to guess another's rule? Do they use feedback from a previous guess when making subsequent guesses?

If you find students who tend to sort in ways that are not consistent, you may want to limit their sorting materials to about 5–8 objects. You might also spend extra time with these students, creating groupings for them to guess, or encourage them to use the Shape Rule Cards.

Collecting Data

- Are students following through on their data plan? How are they collecting responses?
- How are students keeping track of who has already answered their question? Are they accurate?
- Do students understand their data question well enough to explain it to others?

At the end of each session, students will need to clean up their materials. During Session 4, make sure that any students who have not yet collected their survey data have a chance to do so.

Note: After Session 4, make a second copy of each pair's collected data, so that every student has a copy to use during Sessions 5 and 6. If they have used cubes, students will need a duplicate set.

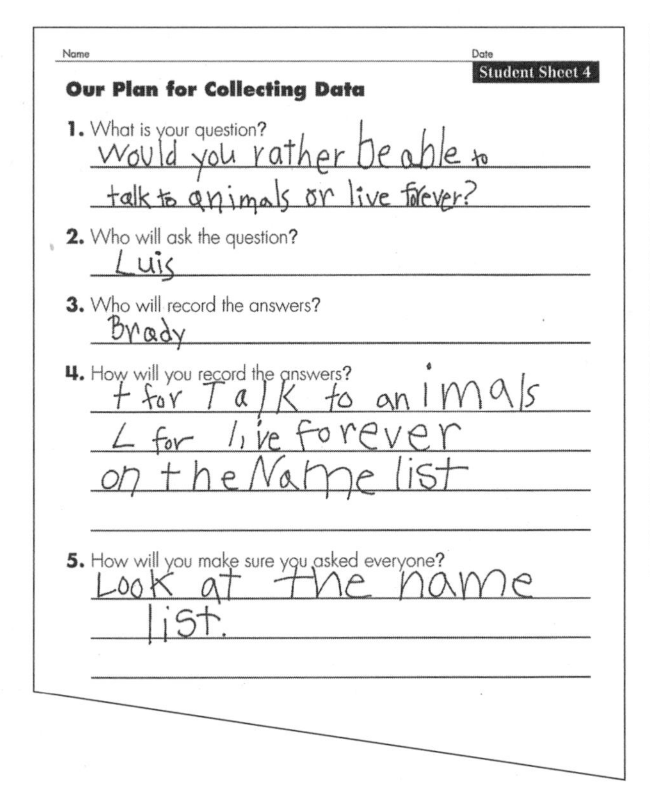
Name Date

Student Sheet 4

Our Plan for Collecting Data

1. What is your question?
Would you rather be able to talk to animals or live forever?

2. Who will ask the question?
Luis

3. Who will record the answers?
Brady

4. How will you record the answers?
t for Talk to animals
L for live forever
on the Name list

5. How will you make sure you asked everyone?
Look at the name list.

Andre L	Chanthou L
Brady L	Claire T
Chris L	Eva L
Diego L	Iris L
Donte L	Jacinta T
Fernando L	Kaneisha L
Garrett L	KristiAnn L
Jamaar L	Leah L
Jonah L	Libby L
Luis L	Mia T
Max L	Michelle T
Nathan L	Nadia L
Tony L	Shavonne L
Tuan L	Susanna L
Yanni L	Tamika L
	Yukiko T

Sessions 3 and 4 Follow-Up

Extension

Sorting with Tabletop, Jr. Tabletop, Jr., software for IBM or Macintosh computers, offers additional sorting activities for students in grades K–3. Data sets include, among others, attribute blocks, party hats, and creatures called Snoods. Of the many games, the following are especially relevant to this investigation:

Guess My Loop: 1

Guess My Loop: NOT

Test My Loop: 1

Test My Loop: NOT

Do Not! Please Do!

If you have a classroom computer, students might use the Tabletop, Jr. software during Choice Time in Sessions 3 and 4. If you introduce the software, you will need additional class time to demonstrate its use, as well as an extra session for Choice Time.

Sessions 5 and 6

Representing and Sharing Survey Results

Materials

- Copies of data from previous sessions
- Large paper (1 sheet per student)
- Stick-on notes, cubes, squares of colored paper, dot stickers
- Markers, glue sticks
- Student Sheet 6 (1 per pair)

What Happens

Using the data they collected during Choice Time, students create a representation of the results of their surveys. Each pair then makes an oral presentation of their findings to the whole group. Student work focuses on:

- inventing representations of data to communicate findings
- explaining and interpreting results of a survey to an audience

Activity

Teacher Checkpoint

Making Representations of Our Data

Return to each pair their collected data (from Choice Time) and the copy you made, so that each student has an individual copy to work with.

Look at the results of your survey and think about what you found out. What are some ways you could show other people what you found out?

You might make representations, like the ones you made for the eagles and whales question. Or, you could invent another way to show how people answered your question.

Encourage students to describe how they might use cubes, stick-on notes, paper and markers, or any other ideas they are considering.

Whatever you decide to do for your representation, be sure to show your findings clearly. Think of these sheets as "telling the story" of what you found out.

Even though you collected your data with a partner, you will each make a representation by yourself. You and your partner might have completely different ideas, and that's great. Then we can see the same data shown in two different ways.

Students work individually for the rest of the class period. They can make more than one representation if there is time. At least one representation should be a permanent recording on paper. Those who choose to use materials such as cubes should also make a drawing of their cube data.

Observing the Students

Here are some things to look for as you observe students working:

- How accurate are they in their counts of the survey data in each category? Have all the data been accounted for in their representation?
- Are students grouping the data that belong together?
- Does the representation communicate the relative size of each category?
- What aspects of their representation are clear? What aspects need further clarification?

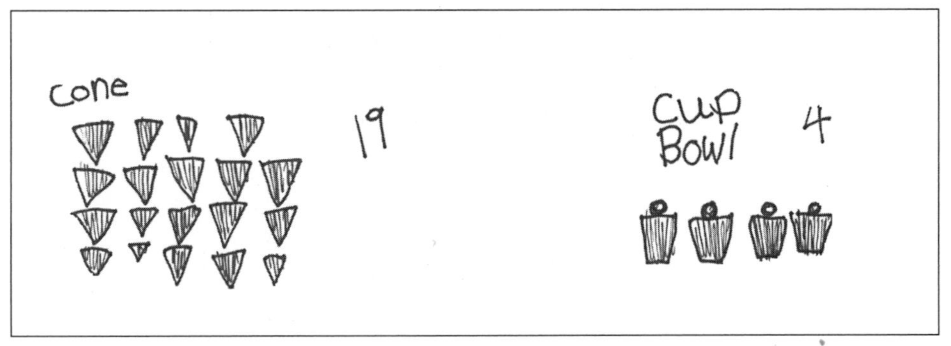

Survey question: Would you rather eat ice cream in a cup or in a cone?

Survey question: Would you rather be invisible or be able to fly?

Activity

Our Findings

As students finish their representations at the end of Session 5 or the beginning of Session 6, distribute Student Sheet 6, Our Findings, one to each pair. This sheet is an opportunity for students to begin thinking about the oral presentations they will be making to report their survey results to the rest of the class.

After you finish representing your data, there's one more important step: communicating what you found to someone else. You'll do that by standing up and telling us about your survey question and the answers you got. You can use this sheet to get ready for your presentation.

Read aloud the three questions on Student Sheet 6:

1. **What was your question?**
2. **What did you find out?**
3. **What surprised or interested you?**

First grade students will generally be able to *say* more about these questions than they will be able to write. Support them as you do for other written work, and remind them that the important thing is to have an answer for these three questions that they can tell to the class. While they are working on this sheet, circulate to be sure they understand the questions and to talk to them about their findings.

When you know the answers to these questions, you can practice presenting your findings to another pair of students. At the end of class, people who are ready may share with the whole group.

You may want to model a way to present findings, using one student's representation of the class findings on the whales and eagles question. For example:

I'm going to show you how a presentation might go. Listen for the three things I'm going to tell you.

- **My question was, "Would you rather be a whale or an eagle?"**
- **I found out that 12 kids wanted to be eagles, 15 kids wanted to be whales, and 2 couldn't decide.** *[Show the representation.]*
- **This was interesting to me because the numbers were pretty close, 15 and 12. I was surprised because more kids wanted to be whales. I thought that more kids would want to be eagles so they could fly, since a lot of kids already know how to swim.**

As students seem ready, match up pairs to practice their presentations. As you circulate and listen to students presenting their ideas, ask questions and give suggestions to help them make clear and confident presentations.

Activity

Presenting to the Group

During the last 20 minutes, students who are willing and prepared make a presentation to the class from the front of the room. Presenters need to bring their representation of the data, and they may bring their sheet of findings as a prompt. When both students in a pair want to present their results, they should do so together.

If this is the students' first oral presentation, explain that they need to speak loudly, look at the audience, explain things clearly, and show their work so that everyone can see. The audience should listen carefully and raise their hands if they have questions. The **Dialogue Box,** Sharing Survey Findings (p. 53), demonstrates how one teacher helped two pairs through their presentations.

You may not have time for everyone to share in one session, nor may everyone be willing to share. If there are more students who want to present, plan to continue another day.

You might also post students' work in the classroom or hallway. Encourage students and parents to browse in this exhibit area.

Sessions 5 and 6 Follow-Up

Homework

Find Your Birth Date For the first session of Investigation 3, Birthdays, students will need to know their birth dates. Even if you already have this information, students will be more involved in the investigation if they bring you this information themselves.

If students in your class do not celebrate birthdays, they may choose another special date. See the **Teacher Note,** Birthday Customs and Celebrations (p. 61), for other suggestions.

Ask students, for their homework, to write down their birth date or other special date and bring it in for the next math class.

Extension

Other Surveys Two of the listings for Related Children's Literature (p. I-17) have survey tie-ins. *The Birth-Order Blues* by Joan Drescher provides an opportunity to compare the survey techniques in the book with those your students have been using. *This Is the Way We Eat Our Lunch* by Edith Baer could lead to a lunch survey in your class. For example: Did you bring lunch today or eat a cafeteria lunch? Would you rather have a hot or cold lunch? For lunch today, did you have a sandwich or something else?

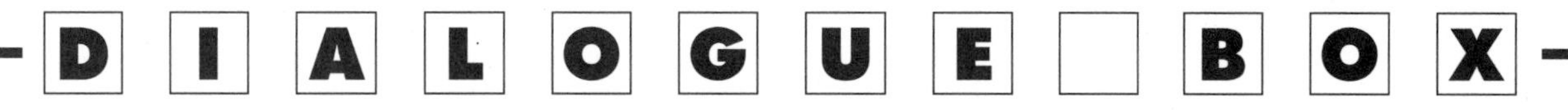

Sharing Survey Findings

These students are presenting their findings from the survey questions they asked each other during Investigation 2. In this class, the teacher's questions guide students through their presentations. In other classes, the teacher may be less involved, depending on the students' comfort and experience with data and with sharing in front of a group. In the following presentations, students focus more on their survey results than on their representations.

Chris and Yanni *[reading together]:* Would you rather talk to animals or live forever?

What did kids say?

Chris: That 22 people want to live forever, and 4 wanted to talk to animals.

Are those numbers close together or is one a lot more?

Chris and Yanni: A *lot* more.

Why do you think more kids wanted to live forever?

Yanni: Because they don't want to die.

Chris: Some people are afraid to die.

I guess if you live forever, you wouldn't have to worry about that. I guess not a lot of people think it's important to be able to talk to animals.

Yanni: If you talk to animals, then maybe you can fly and talk to birds.

Thank you for your report. Now, Claire and Tamika, let's hear from you. What was your question?

Claire: It was about eating your ice cream in a cup or in a cone.

Before you tell us, what do other people predict will be the results of their survey?

Iris: I think more people like ice cream in a cone, because I do and my brother does.

Tony: Yeah, I think so too, because a lot of kids here eat ice cream that way in the cafeteria.

So tell us what they said.

Tamika: Well, 5 people wanted ice cream in a cup and 20 people wanted it in a cone.

Which number is bigger?

Tamika: Twenty is a lot bigger.

Why do you think more people wanted ice cream in a cone?

Claire: Because 5 people need a spoon for the cup and 20 people don't need any spoons.

How do you eat when you eat a cone?

Iris: You just eat it right off. You hold the cone and use your mouth, but then you have to chew the cone.

What do other people think?

Fernando: Some kids like to eat the cone part.

That would be another interesting survey to take. What could we ask if we wanted to know more about that?

Libby: Do you like ice cream cones?

Diego: Or you could ask, "Why do you like to eat ice cream?"

What would your choices be?

Diego: Because of the cone, or because of something else....

Toward the end of this discussion, building on student interest, the teacher brings up another survey question that students might like to explore. This is one good way to extend this activity.

Birthdays

What Happens

Session 1: When Is Your Birthday? Students talk about what a birthday is and figure out how many of them have birthdays in each month. They look at calendars and make cards showing their birth dates, then organize themselves in groups according to their birth month and make a corresponding display of their birth date cards.

Session 2: Whose Birthday Comes Next? Student pairs try different ways of organizing the class birthday data. Then, working as a whole group, they organize the class display by months and by dates within the month. After a discussion of what the completed chart conveys, they put themselves in a "birthday order" lineup across the classroom.

Session 3: Timeline (Excursion) Students listen to the month-by-month story told in *A House for Hermit Crab* by Eric Carle. They then create a group timeline to tell a similar story over one year, drawing a picture of what happens during each month and displaying the results in order.

Routines Refer to the section About Classroom Routines (pp. 100–107) for suggestions on integrating into the school day regular practice of mathematical skills in counting, exploring data, and understanding time and changes.

Mathematical Emphasis

- Becoming familiar with calendar features
- Grouping and describing data about birthdays
- Ordering data about birthdays
- Observing the cyclical nature of the sequence of months

What to Plan Ahead of Time

Materials

- Index cards or half sheets of paper (about 5 by 8 inches): 2 per student and some extras (Sessions 1–2)
- Small envelopes: 1 per pair (Session 2)
- Dark crayons or markers (Session 2)
- *A House for Hermit Crab* by Eric Carle (Scholastic, 1987) (Excursion, Session 3)
- Chart paper (Session 3)
- Large drawing paper: 13 sheets (Session 3)

Other Preparation

- Gather a variety of calendars, both current and older, for students to explore. (Session 1)
- Before Session 1, all students should know when their birthday or other "special day" is. Use class records if some students forget to bring in the date as homework. The **Teacher Note,** Birthday Customs and Celebrations (p. 61), suggests ways to include students who do not celebrate birthdays.
- Use large index cards or a half sheet of paper to prepare a set of birth-date cards, 1 per student. Print the month at the top in large letters, and draw blanks for the student's name and birth date (see example, p. 58). Include one or two extra cards for each month, and for any month in which no one has a birthday. (Session 1)
- Plan how you will display the birth-date cards, grouped by months. If you already have another birthday display of some sort in your classroom, cover it temporarily. The cards can be tacked on a bulletin board, clipped on a clothesline, taped to the chalkboard, or backed with Velcro. The method needs to be flexible so that students can reorder the cards. (Sessions 1–2)
- After Session 1, when the class birth dates have been assembled, prepare a master copy of this data on the Birthday Grid (p. 122). Duplicate and cut apart a set for each student, stored in small envelopes.
- Duplicate Student Sheet 7, Calendars (p. 121), 1 per student, for Session 1 homework. If you have Student Activity Booklets, no copying is needed.
- If you will be doing Excursion Session 3, label 12 sheets of drawing paper with the month names in large, dark letters. Have another sheet available to label as the thirteenth month (same as the first; this month will vary). Find a space to post these 13 sheets side by side as a timeline.

Session 1

When Is Your Birthday?

Materials

- Student birth dates (from homework or school records)
- Birth-date cards (1 per student plus extras)
- Student Sheet 7 (1 per student, homework)
- Birthday Grid (1 master copy)

What Happens

Students talk about what a birthday is and figure out how many of them have birthdays in each month. They look at calendars and make cards showing their birth dates, then organize themselves in groups according to their birth month and make a corresponding display of their birth date cards. Student work focuses on:

- Being resourceful in finding information
- Comparing sizes of groups
- Describing data

Activity

When Is Your Birthday?

As we've been taking surveys in class, we've learned some things about each other by asking questions. Now we're going to collect another piece of information about each one of you: the date of your birthday or another day that is special for you. First tell me what you know about what a birthday is.

Initially, students are likely to talk about what they do on their birthdays to celebrate. For example, "It's when you have a birthday cake… When you invite kids over and you get to play outside…" Another question can prompt them to think about what defines a birthday.

So a lot of you have cake on your birthday, but why is that day special? What is a birthday?

Encourage everyone to share their ideas. You may hear responses like these: "It's when you get older… It's when you are born… Every time you have a birthday you get older and older and older and older."

List all ideas on the board in short form (for example: *birthday cake, party, the day you were born, older and older*). If you have students who have chosen other special days, also talk about the significance of those days.

❖ **Tip for the Linguistically Diverse Classroom** To help students understand the topic of this investigation (birthdays), show pictures of birthday celebrations, display birthday cards, and sing birthday songs that reflect the traditions of students in your class.

We're going to be asking exactly when each of you has a birthday. In our class, which months do you think will have the most birthdays? Do you think there are months with *no* birthdays?

Students share their predictions, which will likely be based on their own birthdays and their knowledge of friends' birthdays.

Who knows what month their birthday is in?

Ask a few students to give their birth dates. Write the months on the board or on chart paper in the order they are mentioned. If students are unsure, they may consult their homework or your school records.

Let's see, we have April, October, and July. Who has a birthday in a different month?

List as many months as students name.

Do we have all the months here? How many months are there in a year? Which ones are missing?

Students may suggest that you write the months with no class birthdays in a different location. Follow their suggestions for writing all the months; these are not likely to be in order. If students cannot think of all 12, leave room on the list for the "mystery months." If students do not know that there are 12 months, or if there are differing opinions about the total number, list on the board all their ideas about how many months there are. Make a separate list of the mystery questions that need to be solved; for example, students in one class wondered, "Is summer a month?"

How could you find out for sure how many months there are? How could you find out which months are missing? How could you figure out the order of the months?

See the **Teacher Note,** Finding Out For Yourself (p. 62), for information about encouraging student resourcefulness. Students may suggest looking at a calendar, checking in a book, or finding a copy of the well-known rhyme, "Thirty days hath September...." Make available your collection of calendars and any other resources for students to explore in pairs.

As you look through these calendars, think about the questions we have listed. You will have 5 or 10 minutes to look at a couple of different calendars. (You can trade with someone after you have finished looking at yours.) Then I'll ask you to share the information you have discovered.

Encourage students to solve any of your class mysteries about the months (such as How many months are there? What months are missing in our birthday display?), as well as any other calendar questions they have.

After about 10 minutes, ask students to report on what they found out from the calendars. Record on chart paper what they report. They may share new discoveries as well as answers to the original list of mystery questions.

Keep these calendars available for reference during this investigation.

Activity

How Many in Each Month?

On a table or other large surface, set out the birth-date cards you have prepared. Provide a separate stack of cards for each month, in any order. Students come up in pairs or threes and select their birth month from the stacks of cards. They fill in their birth date and write their name on their card.

January 12
Eva

Now let's find out how many of you have birthdays in each month. We need to know how many are in October, how many are in June, and so on, for each month. Let's try having students with the same month find each other and make a group.

There may be a bit of initial confusion, but students enjoy the involvement of the physical activity and are usually able to form their groups successfully. You may want to help groups find a location in the room. Once the groups are formed, ask students to double-check.

How could you make sure that everyone for your month is in your group?

Some might suggest sending out scouts to check other groups. Others may suggest that they all hold their cards over their heads, then look around the classroom to check. Sometimes two groups will be formed for the same month, but students usually discover this and correct it by themselves.

Now that you have formed your groups and checked to make sure everyone's there, tell me what you can see by looking around the room.

Record their ideas on chart paper or on the board. They may notice, for example, that "I'm the only person in my group," that "the biggest group is by the windows," or that "there are four for May." Encourage students to use language that describes comparisons: *about the same, bigger than* or *smaller than, more* or *less.*

Which month has the most birthdays? What's another big birthday month? Are there any months when no one has a birthday?

To start a display of this data, students bring up their birth-date cards, one group (month) at a time. You might tack the cards to a bulletin board or clip them on a clothesline. At this time, make no attempt to organize them; put them up in the order you get them from the students (which may or may not be in the usual order of the months). If students object, explain that they will be working to organize the birthday data during the next session.

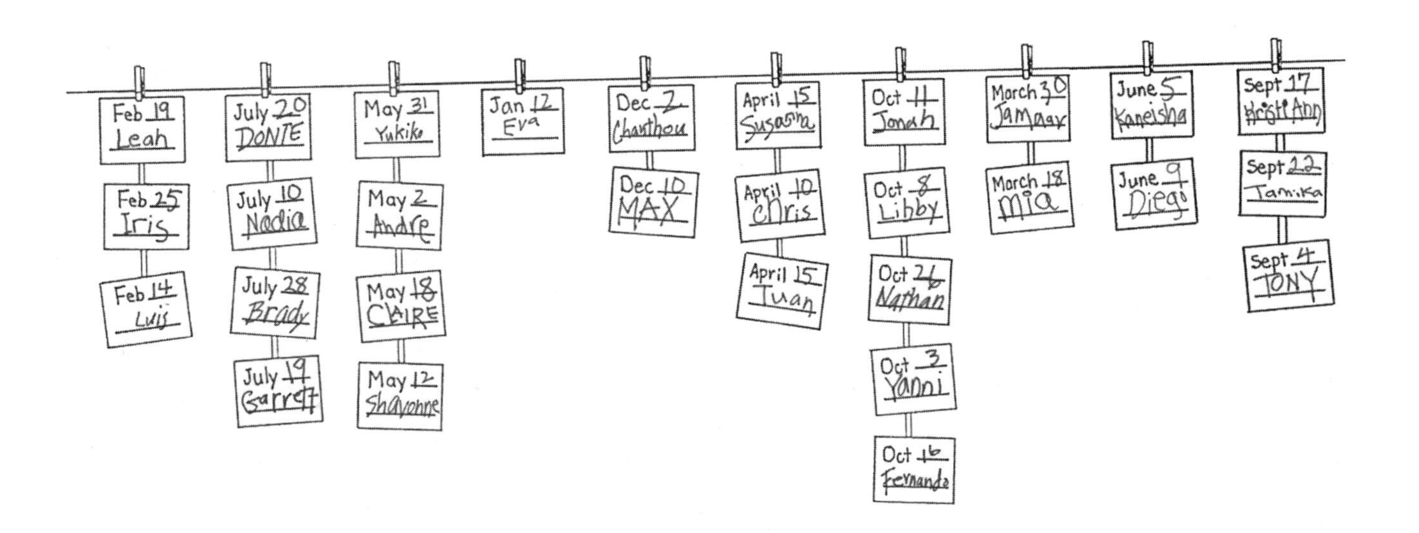

Note: Before the next session, you will need to copy the information from students' completed birth-date cards onto the Birthday Grid (p. 122). Duplicate a set of the data for each pair of students. Cut the grid squares apart and put each pair's complete set of class data in an envelope for use in Session 2.

Session 1 Follow-Up

Calendars Students take home Student Sheet 7, Calendars. If they are unlikely to have calendars available, consider sending home old calendars, or a few duplicated calendar pages, for use with this sheet. Students write at least three things that they discover about the days and months.

❖ **Tip for the Linguistically Diverse Classroom** Students might mark directly on calendar pages to show their discoveries. In addition, families can help with Student Sheet 7 in their native language.

Birthday Customs and Celebrations

Teacher Note

In connection with this birthday investigation, you might plan some special celebrations. Keep in mind that some cultural and religious groups de-emphasize individual birthdays. Some families may feel comfortable with their children celebrating a special day that is not their birthday—for example, another day that is important for them as an individual or family. At the beginning of these activities, keep discussion open so that students can talk freely about particular customs that their families follow. In addition, consider calling these "special days" instead of "birthday celebrations," so that any students who do *not* celebrate birthdays can still feel part of the festivities.

Making decisions with your students about how to celebrate can involve some good mathematical problem solving. Consider these two questions as you plan your celebrations:

- When will we celebrate everyone's special days?
- What "special treat" will we have for the celebration?

Interesting discussion can grow out of these two questions. Start by focusing on one of the months with several birthdays. Some classes decide to celebrate on each student's birthday; others choose one day as the special day for all the students born in that month. Other classes do a combination of special days and individual birthday celebrations.

If you or the class decides to limit the special day to one a month, how will that day be chosen? Should it be on the first birthday that comes up that month? the last? the middle birthday? A day exactly in the middle of the month? How can you figure out where the exact middle of the month is? This is a nice problem for pairs of students to work on, using a page from a calendar.

Another part of the discussion will focus on what to do about students with summer birthdays and/or students with birthdays in months that have already passed for this school year. Probably students will bring up this issue themselves; if not, introduce it. These questions often provoke lively talk about what to do. Student ideas have included suggestions like these:

- Put all the summer people into February because no one has a birthday in February.
- January, March, and May only have one birthday each, so summer people celebrate with them.
- The summer people get to choose which month they want.

After this decision is made, you might make a second chart or mark a calendar with your students, showing the dates chosen for the celebrations and the names of the students to be honored on those days.

Finally, students plan what the special treats will be. You will set the guidelines for this discussion, given the customs and policies of your school. One teacher set these guidelines: the treats could not involve food, and could not take away too much learning time.

After brainstorming a list of possible ideas (such as extra play time, story time outside, go to the playground, take a walk, make masks, read books for fun), each birthday group could meet together to choose their treat. Then help the class list the treats for the special day in each month.

Teacher Note — *Finding Out for Yourself*

Young students have a lot of authorities in their lives, from whom they get a great deal of information and advice. However, even first graders can and should learn how to find information for themselves, using their own resources. Your students may believe that the only way to obtain factual information is to ask someone for it. You can encourage them to develop other strategies as well—that is what collecting data is all about.

To help students become resourceful seekers of information, you must be particularly alert in identifying questions to which they can find their own answers. For example, when a student asks, "What month comes after February?" it is tempting to reply simply, "March." If, instead, you suggest that the two of you go over to the wall calendar to find out, you will be empowering that student by adding to his or her resources.

It is certainly not possible, or desirable, to send students out to find their own answers for every question they have. For example, you don't want to send students to a dictionary every time they ask how to spell a word. However, if someone is always willing to spell the word for them, students will develop no resources on their own. A better alternative is to encourage students to rely on everything they know about letter symbols and their sounds, rather than immediately turn to adult authority. In some situations, you might accept students' "invented spelling." Or you might help them set up their own word files that they can use directly.

Similarly, it is not always appropriate in mathematics to say, "Figure it out for yourself"; nor is it always appropriate simply to supply the answer. You can help your students expand their repertoire of strategies for tackling unfamiliar problems by providing a concrete material with which they can model the problem, by reminding them about another problem they solved, by suggesting collaboration with another students, or by encouraging them to collect their own data.

Helping students become resourceful will enable them to become more independent—not only during their data experiences, but throughout their mathematics work.

Session 2

Whose Birthday Comes Next?

What Happens

Student pairs try different ways of organizing the class birthday data. Then, working as a whole group, they organize the class display by months and by dates within the month. After a discussion of what the completed chart conveys, they put themselves in a "birthday order" lineup across the classroom. Student work focuses on:

- organizing and sequencing data
- interpreting data
- observing the cyclical nature of the months

Materials

- Class sets of birthday data, in envelopes (1 set per pair)
- Index cards (large) or half sheets of paper (1 per student)
- Dark crayons or markers

Activity

Organizing the Birthday Data

Focus attention on the class display students made in the last session, showing birth dates by month.

Suppose we had a visitor to our class. What could that visitor find out about us from this chart?

Students are first likely to point out information about individuals; for example when their own birthday is. They may also begin to generalize information about how many birthdays occur in different months. Also ask students specific questions about your data set.

Whose birthday is in March? What is the date of her birthday? Which two children have birthdays in the same month? What month has the most birthdays?

As children become familiar with the chart, ask them to pose their own questions that they think can be answered by looking at the chart. See the **Dialogue Box,** Looking at the Birthday Chart (p. 68), for examples of students' ideas.

When we first put up our birthday data, we didn't try to organize it, except in groups by month. Today, we'll do more organizing. There are lots of different ways we might organize the chart. In a few moments we will decide together how we want to do that. But first, you and your partner are going to try out your own ideas.

I'm giving each pair a small set of our class birthday data. You'll find it all in this envelope. Be careful not to lose any pieces. See what kind of groups you can make, and decide on the order you think they should go in. Which month should we start with? You decide that with your partner.

Distribute an envelope of class birthday data to each pair of students. Pairs work together to arrange their data.

Observing the Students

You may discover that some students already know the usual order of the months. These students should focus on which month they'd like to begin with. Other students may spend most of their time becoming familiar with the usual order of the months. As you watch the students working, notice how they are grouping the data.

- Are they putting the cards into groups that go together (by month)?
- Does their ordering make sense in the context of the birthday activity?
- Are they using any resources, such as their homework from Session 1 or classroom calendars?
- How familiar are students with the usual order of months?

Activity

Whose Birthday Comes Next?

Ask students to share some of the ways that they organized their sets of class birthday data. Then ask them to look at the large class display of the same data. Students use their own set of data displays to contribute to this conversation. (If it is too distracting to have their birthday data groupings out during this discussion, you may need to collect these.)

What if we wanted to know whose birthday comes next in the school year, and then the next one, and the next one? How could we set up our display so that we could easily tell?

Students share ideas about what order to put the months in. Sometimes students may suggest alphabetical order of months or numerical order of all the dates. Even though these sequences will not show which month follows which, give students a chance to resolve the issue themselves through discussion, rather than explaining why such an arrangement won't work. The **Teacher Note,** The Cyclical Nature of Time (p. 67), discusses ideas students are likely to encounter during this discussion.

If the chart is not too fragile, students can move some of the cards as they decide on the order—or you may prefer to do it yourself. Some students are likely to feel strongly that the months should be in the usual calendar order, while others will want to try alternative ideas. Even though standard calendar order is the best way to organize this data set, there are decisions to be made about what month it makes sense to begin on. Encourage different points of view about what month to use to start the chart.

Jamaar and Libby want to start with January, like a lot of calendars do. Mia suggested starting with the month we're in. Can anyone think of a different way we could start?

Some classes decide to start with January (the beginning of the calendar year), some with September (the beginning of their school year), and some with the current month. You and your class can decide what makes the most sense for you.

So, how many months are there? Are there any missing?

If there are months missing, students name them and offer suggestions about how to indicate them on the chart. They may also want to add some indication to the chart that there isn't anyone in their class with a birthday in that month.

Call attention to whatever month is first in your display.

What month is right after [September]? What month is right *before* [September]? How can we tell?

Students are likely to look to the right to find the month that comes *after,* but may not know where to look for the month that comes before the first month in their display. They may need to look at other calendars, or name all the months in order to find the sequence.

If students have not yet brought up the problem of ordering the birth dates within each month, ask them explicitly about this.

Now that we have the months in order, how can we tell which birthday is the next one, and which one comes after that?

Give students lots of time to share their ideas about ordering the birthdays. They may need to physically move around all the cards from one month in order to tell what order they go in. As students reorder the chart, ask them to explain how they know the order of the numbers.

Why did you put 18 before 25? Why does the 10 go first in July and last in December?

See the **Dialogue Box,** Whose Birthday Comes Next ? (p. 69), for examples of student thinking. When the chart is finished, ask students once again to describe what a visitor to the class might find out from looking at their chart.

Now that we've organized our data, what can someone tell about the birthdays in our class from this display?

At this point all students should be able to make an observation that involves looking at more than one data point. Students may comment on the order of the months and the order of their birth dates, as well as on what months have the most and least birthdays.

Save and display this birthday chart along with other birthday displays your class may have made previously.

Activity

Birthday Lineup

Distribute index cards or half sheets of paper and ask students to write their birth date (for example, April 28) in large letters, using a dark crayon or marker. Explain that they will use these to put themselves in birthday order.

Note: If you think your students may already have a good sense of the order of their birthdays, the order of the months, and their birthday display, you may want to try this activity with the class birthday display hidden.

Now you're going to put yourselves in birthday order. People in the first month [on the class display] should line up starting on this side of the room. People in the last month will go over there. You will make one big line by yourselves. When you think you are in the right place, you can sit down where you are in line.

Encourage students to do this without adult help. They may use the calendars, any birthday lists or displays in the classroom, and their own sense of the order of the months and the days. Students may call out the names of the months in order, or may figure out which months belong where, then sort within each month by days. When they are settled in an orderly line, ask:

How does your line compare to the order of our birthday data display? How could we check?

As students compare their lineup to the birthday chart, anyone who is in the wrong place may move.

Session 2 Follow-Up

Special Celebrations Make a plan, based on the birthday chart, for special ways of celebrating the birthdays (or other special days) in each month. See the **Teacher Note,** Birthday Customs and Celebrations (p. 61), for more details.

Extension

Teacher Note

The Cyclical Nature of Time

Birthdays, regular daily activities, and the seasons are recurring events or cycles that children use as they begin to make sense of the flow of time. Grasping the idea of a cycle, such as the months of the year, involves having a sense of pattern and sequence.

During this investigation, the cyclical nature of the sequence of months is likely to come up. In one class, for example, the students eventually decided to begin their chart with the current month (November), with the other months following in order. When a month was over, they would move it to the end of the line, so that the current month would always be first and they would know whose birthdays were in that month. As one student observed, "You can't stop the months. They just keep going, going, going, going."

No matter how students decide to start their data display (that is, no matter which month is first), the months will eventually need to be placed in the conventionally accepted order. During this discussion, students will come up against the concepts *before* and *after.* Developing an understanding that August is always before September, even if it is very far away on the data display, is an indication that students have a sense of the predictable nature of the yearly cycle of months.

Excursion Session 3, which involves reading the book *A House for Hermit Crab* by Eric Carle and creating a timeline to go with it, provides another context in which to discuss the cycle of the months. Students will also further explore the cyclical nature of time in the *Investigations* grade 2 unit, *Timelines and Rhythm Patterns.*

DIALOGUE BOX

Looking at the Birthday Chart

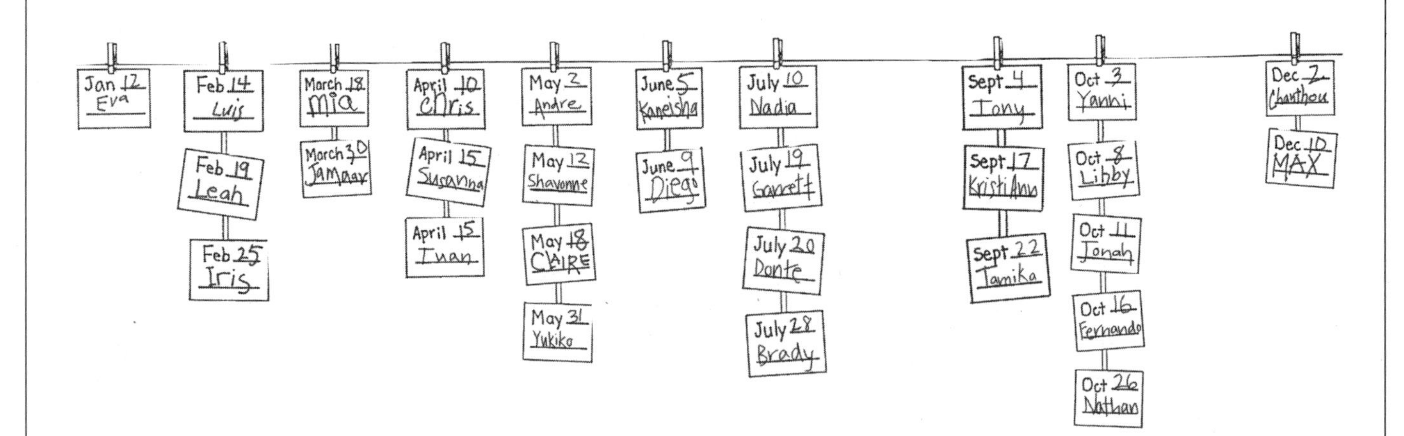

As these students look at their class birthday data chart, they begin to get a sense of how the birthday data are distributed across the months. They notice the months with the least and the most, and even discover the holes in the data set.

So, by looking at this birthday data, what could people tell about the birthdays in our class?

Kaneisha: They could see that my birthday is in June.

Donte: They could tell that Tony's birthday is on September 4.

So is this everyone? How can we be sure we have all the data?

Brady: Count.

How many cards should there be?

Max: 29. Because there are 30 in our class and Jacinta is missing.

[The students count aloud together from 1 to 29 while Andre points to the cards.]

Susanna: There are different numbers on each month. Some have one, some have more.

Claire *[pointing]:* There's three at this one [February], and two at that one [December].

Luis: January is less, it has only one.

Tamika: No, the less is zero, because we don't have all the months.

Which are we missing?

Tamika *[counting]:* We have ten months up here…

How many months would there be if we had them all?

Shavonne: Twelve. So there are one, two missing.

Fernando: We're missing Jacinta's month [August] and November, too.

How could you tell?

Fernando: I looked at the list of the months.

So which month has the least birthdays?

Tamika: November! Because Jacinta's birthday is in August.

What else can you say about the birthdays in each month?

Nathan: October is the biggest. It's the king of the classroom because there's more there than any other month.

Whose Birthday Comes Next?

These students have put the twelve months in order on their class birthday display. Now they are organizing the dates within each month. In this class, the display begins with January, a month with one student's birthday.

How should we order these birthdays? Who's first?

A few students: Eva!

Luis: Then me [February 14]. Then Leah [February 19], then Iris [February 25].

How do you know?

Luis: Because the teens always come before the twenties.

What about March? We have Mia on March 18 and Jamaar on March 30.

Iris: Put 18 first.

How did you decide that Mia is next?

Iris: I looked at the number line.

For April we have three people [Susanna, April 15; Chris, April 10; Tuan, April 15]. **Raise your hand if your month is April.**

Jonah: We have a tie!

Wow! Two people on April 15! Who should we put first?

Shavonne: Susanna.

Of all three people, who goes first?

Nadia: The 10 [Chris].

Fernando: I know what to do. Write the two names [Susanna and Tuan] together.

Andre: Or you could do alphabetical order.

What do you think, Susanna and Tuan?

Susanna: Let's go together.

[The teacher puts their cards together, rather than one above the other.]

Andre: I go next because I'm the second of May.

[The ordering continues in this way through July. Then Yanni notes a month that is missing.]

Yanni: I wish we had August.

Why is that?

Yanni: Because the months always go in order. You never skip a month.

Should we include the names of the months that are missing?

A few students: Yeah!

Kristi Ann: We don't have cards for them, but we'll put the name there [between July and September]. And then when Jacinta comes back, she can do a card for August, 'cause that's when her birthday is.

Session 3 (Excursion)

Timeline

Materials

- *A House for Hermit Crab*
- Twelve sheets of drawing paper, labeled with month names, and one extra
- Crayons or markers
- Chart paper

What Happens

Students listen to the month-by-month story told in *A House for Hermit Crab* by Eric Carle. They then create a group timeline to tell a similar story over one year, drawing a picture of what happens during each month and displaying the results in order. Student work focuses on:

- sequencing data
- noticing the cyclical nature of the months
- creating a timeline representation

Activity

A House for Hermit Crab

Read to the class Eric Carle's book, *A House for Hermit Crab.* This is a story about a hermit crab who has outgrown his shell. He finds another, but it is too plain. Each month of the year he finds another sea creature to help him decorate his shell. For example, in April, a sea star agrees to decorate his house; in May, a piece of coral makes his house more beautiful. One year later, the crab outgrows his lovely shell and must search for another.

As you are reading the book, stop periodically and at the end to discuss the changes that the hermit crab has gone through. For example:

What do you think will happen next month? Does the crab ever lose any of his decorations?

Every month the crab gets more and more decorations, but what happens at the end? What do you think will happen next year?

This story begins and ends with the same month. Given the cyclical nature of time, we can infer that the hermit crab's next year will be similar to his last year.

Now we are going to make up our own story about a hermit crab like the one in the book we just read.

Note: You might develop a different story with your class, for example, about a character who plants new things in a garden each month. In any case, be sure to keep the essential features of the crab's story: the accumulation of things over time, and the cyclical nature of time.

In our story, we'll show what happened, month by month. First we need to make sure we have all the months, and in the right order. What month shall we start with?

Students may suggest January, September, or the current month; any of these is a fine starting place. Ask them to brainstorm the rest of the months in order as you list them, vertically, on chart paper. When students have named all 12 months, ask them what month would come next. Record that as well, so that the last and first month on your list are the same.

Now we need to decide what will happen to our hermit crab every month as she decorates her shell.

Students make suggestions for what happens each month. As in the story, the first month may show an empty shell. Each month after should show one additional decoration, so that they accumulate over the year. Record students' ideas on chart paper next to the corresponding month names. Students ideas may come from the story, from their familiarity with sea creatures, from other books, or from their imaginations.

January sand dollar
February barnacle
March sea anemone
April starfish
May mermaid
June seaweed
July baby octopus
August seahorse
September sting ray
October clownfish
November shark
December sea turtle
January

Activity

Creating a Timeline

After the class has a complete plan for their story, show them the 13 sheets of drawing paper. Explain that they will work in pairs to draw a picture for each month of the story.

When you have finished your drawings, we will post them in order, along one wall, in a *timeline.* A timeline will help us retell the story of what happens as time moves along from month to month to month. Putting these sheets up to make a timeline will allow us to see all the information at once.

Ask students if they have seen timelines before. You might discuss the class birthday chart as being a timeline that represents one year.

Now, our story doesn't really end after 12 months. How can we show on our timeline that these events go on and on?

Possible ideas include making sheets for even more months to show how the story repeats, writing a note to explain what happens next, or adding a symbol (such as an arrow or ellipsis points) to show that the story keeps on going.

Before we decide who does each of the pictures for our timeline, let's think about what you'll need to draw. *[Point to a month on the list that is midway through the story.]* **If you were going to draw the month of [May], what might you show?**

Students are likely to mention just the new decoration shown on the list. If they don't mention it, remind them about the previous months.

Emphasize that students need to show in their illustrations all the accumulated decorations for the year so far. Their picture is a "snapshot in time," and if they draw only the new element for a particular month, they aren't showing the situation as it really is, with all that came before.

Depending on your class size and how you have decided to show that the story continues on and on, your timeline may have 13 or more sheets. With a large class, you might assign two or three students to a single drawing and/or plan to continue the months further into the second year.

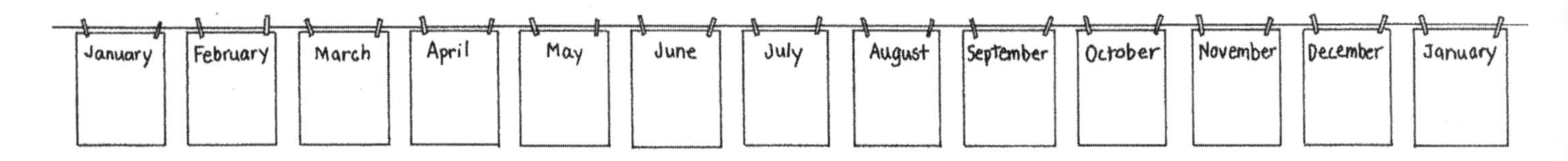

How to display the finished work as a timeline

As students decide what months they want to illustrate, write their initials next to that month on your list. Then distribute the previously labeled sheets of drawing paper to the appropriate pair.

Pairs of students draw a picture of their character's changes for the month that they chose. You may need to remind some students that their drawings should reflect not only the new decorations but those from the previous months as well.

Display the finished timeline sheets on a wall or clothesline, evenly spaced, in order. Students might help you order the illustrations as they are finished. Find a time to retell the class story together, following the timeline.

INVESTIGATION 4

Ages and Attendance

What Happens

Session 1: How Old Are We? Students collect data about their own ages, using a variety of methods to represent, describe, and compare the ages in their class. As homework, they collect information about the ages of people in their families for use in the next session.

Sessions 2 and 3: Representing Data on Family Ages Using the data from their homework, students represent their family members in pictures and put them in order by age. These groupings are used for a guessing game. Finally, the students make a class chart showing the ages of themselves and their siblings, and discuss what they can find out from this data display.

Sessions 4 and 5: Attendance Comparisons As a class, students collect attendance data. They create an "unusual day" scenario and, as an assessment, make their own representation of the attendance situation their story describes. They compare their real attendance data with their unusual day attendance data.

Routines Refer to the section About Classroom Routines (pp. 100–107) for suggestions on integrating into the school day regular practice of mathematical skills in counting, exploring data, and understanding time and changes.

Mathematical Emphasis

- Creating a variety of representations of several categories of data
- Describing data qualitatively ("a lot more here, hardly any absent") and quantitatively ("25 here, 2 absent")
- Interpreting data that shows values (ages) and categories (siblings and selves) at the same time
- Comparing two data sets

What to Plan Ahead of Time

Materials

- Interlocking cubes: class set (all sessions)
- Scissors: 1 per pair (Session 2)
- Stick-on notes in two colors: about 6 per student (Sessions 2–3)
- Scratch paper (Sessions 2–3)
- Crayons or markers (Sessions 2–5)
- Chart paper (Sessions 2–5)
- Construction paper in two colors: several sheets (Sessions 4–5)
- Large paper (11 by 17 inches): 1 sheet per student (Sessions 4–5)
- Materials for making representations, such as stick-on notes, dot stickers (Sessions 4–5, optional)
- Glue sticks to share (Sessions 4–5)

Other Preparation

- Duplicate the following student sheets and teaching resources, located at the end of this unit. If you have Student Activity Booklets, copy only the extra sheets marked with an asterisk.

  For Session 1
  Student Sheet 8, Record of Family Ages (p. 123): 1 per student, homework

  For Sessions 2 and 3
  Student Sheet 9, Family Portraits (p. 124): 1 per student, plus extras* (for large families)

  100 Chart (p. 125): 1 per student

- Cut rectangles about 2 by 3 inches from construction paper in two colors. Make two of each color for every student, plus some extras. (Sessions 4–5)
- Duplicate your class list, 1 per student and some extras. (Sessions 4–5)
- Prepare an Attendance Chart by taping together two sheets of chart paper, one above the other, and posting them so the bottom is nearly touching the floor. (Sessions 4–5)

Session 1

How Old Are We?

Materials

- Interlocking cubes (class set)
- Student Sheet 8 (1 per student)

What Happens

Students collect data about their own ages, using a variety of methods to represent, describe, and compare the ages in their class. As homework, they collect information about the ages of people in their families for use in the next session. Student work focuses on:

- comparing several categories of data through questions that begin "How many more" or "How many less"
- categorizing data so that the number of items at each value can be compared
- representing information about several categories
- using different representations, such as cube towers and tallies, that help in comparing categories
- using numbers that stand for quantities that cannot be physically counted (such as ages)

Activity

How Old Are We?

Today we are going to talk about all the different ages of the children in our class. What do you think is the youngest age of someone in this class? What would the oldest age probably be?

Take students' suggestions and write a list of possible ages on the board, for example:

5 6 7 8

How could we show how many children in our class are each of these different ages?

You might hear a number of suggestions, for example:

- Take a show of hands for each age and record the total for each group.
- Write each person's name below the corresponding age and count the names.
- Have everyone who is a certain age stand in a different line. Someone can take a count of everyone in that line and write the total for that group next to the right age.
- Each student writes his or her name on the board with the age after it.

Try two or three of students' suggestions to demonstrate that there are different ways of collecting and displaying the same information. If students don't think of more than one possibility, use an idea from above or another of your own. Describe each as a different way of *representing* the data about ages. Ask students to check if their results are the same each time.

Tallies If no one in the class mentions tallies, tell the class you want to show them a method some people use to quickly collect and record data. Go around the class in some order (perhaps by tables or by rows), asking each student to say his or her age. For each student, make a tally mark next to the correct age on the board. Point out that this is how you are keeping track of the data.

For every fifth piece of data for the same age, make a diagonal line across the previous group of four tallies, and point out that this makes groups of five that are easier to count later.

When you have finished collecting all the data, ask students to help you figure out how many are represented in each row. For example, if you have 12 students who are 6 years old, some students might count the tally marks by 1's, while others might recognize the two 5's and count on this way: "5 and 5 is 10, and then two more is 11, 12." If students count by 1's, they will need to remember to count the diagonal line on each group of 5, as well as the 4 vertical marks. Some students may readily use the 5's as a group, just as they use the fingers on one hand to represent 5.

Cube Representations Distribute a handful of about 8–10 cubes to each student.

You're going to use these cubes to make another representation of your ages. First, build a tower to show how old you are. For example, if you are 6 years old, your cube tower should have 6 cubes in it.

After the students have built cube towers for their ages, ask the youngest age group in the class to line up their towers on a table or along the chalkboard tray. Do the same with each successive age group.

When all the towers are lined up, ask students what they can tell about the ages in the class from looking at the towers. Students are likely to notice how many students there are at each age level. Again, check to see if these numbers agree with the results they got previously. Someone may notice that you can tell how many students there are in class by counting all the towers.

Explore questions like these:

How old are most of the students in our class? How many students are that age?

Are there more 7-year-olds in this class than 6-year-olds? A lot more? How many more?

Are there fewer 5-year-olds than 7-year-olds? How many fewer?

Note: Conceptualizing the question "How many more?" takes time. If students have difficulty comparing the categories of data using the cube representations, try having the students themselves line up, in parallel lines, according to the two different ages you are comparing. They can then pair up across the lines, by holding someone's hand or reaching toward them, to get a sense of who does not have a partner, and thus how much longer one line is than the other. Encourage student thinking about the comparison this way:

Why don't these students have partners? What does that mean about the number of 7-year-olds? How many students don't have partners?

Data Changing Over Time As you finish this activity, encourage students to look back over the various representations and think about these questions:

How would our representations of ages in this class be different if we looked at them one month (or two months) from today? What would change? What would stay the same?

Encourage students to discuss their theories about what changes there might be. Students may want to use the birthday chart they made in the previous investigation as they think about this issue.

Activity

Listing Family Ages

Distribute Student Sheet 8, Record of Family Ages, to discuss today's homework.

Today we talked about the ages of the children in our class—your ages. Tomorrow, we're going to talk about the ages of people in our families. So, for homework, you are going to be collecting that information. You need to find out how old everyone in your family is.

Does anyone have a younger brother or sister? What about an older brother or sister? What do you suppose is the youngest age we might find? Any ideas about the oldest we might find? About how old do you think most mothers are?

Students contribute some examples and speculate a bit about the information they will be gathering.

Note: Exactly who to consider "a family member" need not be an issue for students during this activity. For example, you might decide that in their immediate family, students include mother, father, guardians, brothers, and sisters, even if they do not all live in the same home. Students might also include more distant relatives who live with them, such as a grandmother or cousin (but only people, not pets).

This is a good time to introduce the word *sibling,* which sounds important and is appealing to first graders.

Some of the ages you'll be getting are for your siblings. Your siblings are your brothers and sisters. For example, if you have one brother and one sister, you have two siblings. Who has even more siblings?

Session 1 Follow-Up

Collecting Family Age Data Students collect age data for their family members on Student Sheet 8, Record of Family Ages. Emphasize that they should bring their data in for the next class session. Be sure students understand that they can include as family members a person who doesn't live with them all the time (for example, a father who lives somewhere else and is visited only in the summer).

After you distribute the student sheet, point out that there are lines for as many as eight family members, but that students will fill in only the lines they need. They should understand that it's perfectly fine if they need only two lines, for example.

Students will need their homework in order to proceed with Session 2. Prepare a back-up plan for students who forget, such as asking them to fill in another sheet, listing the ages they know and estimating any they don't know for sure.

Sessions 2 and 3

Representing Data on Family Ages

Materials

- Completed homework (family ages)
- Crayons or markers to share
- Student Sheet 9 (1 per student, plus extras)
- 100 charts (1 per student)
- Scissors (1 per student)
- Scratch paper (2–3 slips per student)
- Chart paper
- Stick-on notes in two colors (small)

What Happens

Using the data from their homework, students represent their family members in pictures and put them in order by age. These groupings are used for a guessing game. Finally, the students make a class chart showing the ages of themselves and their siblings, and discuss what they can find out from this data display. Student work focuses on:

- categorizing data so that the number of items at each value can be compared
- representing information about several categories
- using different representations, such as cube towers and tallies, that allow students to compare categories

Activity

Family Portraits

For this activity students need their completed homework on Student Sheet 8, Record of Family Ages. Also hand out Student Sheet 9, Family Portraits. There are six cards on the sheet, so provide extra copies to students who have more than six people in their families.

Today we'll be playing a guessing game about our families. Before we can play, everyone needs to make drawings of the people in their families. Draw one of your family members in each card on this sheet. Don't forget to include yourself!

Label each portrait with the person's name. In the little box at the top of each card, where it says "Age," write that person's age. You can use your homework sheet to help you.

Be sure to put only one family member (and that person's age and name) in each of these cards, because we are going to cut them apart to make a separate card for each person. When you have drawn and labeled everyone, cut on the dotted lines to make six big cards. Write your name on the back of each card.

You might draw an example on the board to clarify the directions.

As students make their pictures, encourage them to talk to each other informally, comparing ages of the people they are drawing: Who is older? Who is younger? Is anyone the same age as someone in your family?

Ordering Families by Age As students finish drawing and cutting apart their portraits, they arrange their cards in front of them so the family members are in age order (either youngest to oldest or oldest to youngest).

For this part of the activity, distribute the 100 charts for students to use as a resource. If students are more familiar with number lines, be sure these are available instead.

Don't expect every student to know exactly where to look for each number on the 100 chart or number line. Some students will be able to find any number quickly, while others will look for each number randomly. Some students will count from the beginning, while others will know approximately where they might begin looking. Some students may use the strategy of reading the numbers aloud in order, listening for a number that matches an age in their family.

If students are still having trouble ordering the older ages, you might suggest that they focus on siblings only.

Activity

Ordering People by Age

When students all have their own family members ordered by age, put the class in groups of three, being sure that there's not an exceptionally large or small number of family members in any particular group. Most groups of three will have between 9 and 15 family members represented. If one group has very small families, you might add a fourth person.

Now we're going to get ready for our guessing game. You'll be playing the game in groups of three. To set up the game, you need to put together *all* the family portrait cards in one lineup, and arrange them in order by age. Line them up with the youngest person at one end and the oldest at the other end. Then put everyone else in between, so the ages go in order from the smallest number to the largest number. You can use the 100 chart [or number lines] to help you figure out which numbers are bigger than others.

To model this process, give an example of how you might order the ages of a few people using the 100 chart as a resource. Or, show the class how one student ordered the cards for his or her family.

Students spend about 10 minutes arranging their group's family portraits by age. Encourage those having difficulty to use a 100 chart or a number line, or to ask each other for help.

When a group has cards for people who are the same age, students will need to decide how to handle these. One approach is to group all the same-age people together; another approach is to decide who among the 6-year-olds is *really* closer in age to 7. Although this second approach is more challenging, students often enjoy getting specific about ordering their own ages.

As students are working, notice what resources they use to order the family portrait cards.

Who's the Mystery Person?

When groups have their family portraits in a lineup, arranged in order by age, where everyone in the group can see them, introduce the guessing game.

Here's how to play a guessing game with your family portraits. One of you starts by choosing a "mystery person" from the portraits in your group. Keep your mystery person a secret. The rest of the group then tries to figure out who your mystery person is by asking you questions. Here's an important part of the game: The guessing questions have to compare two of the cards. I'll do an example to show you how it works.

Quickly sketch a group of about six family portrait cards on the board.

OK, I've picked a mystery person from this family, and you get to guess who it is. But you can't just guess a particular person and ask, "Is it Kayla?" or "Is it 6?" You can only ask questions that *compare* two cards—questions like these:

- **Is the mystery person *older than* Steven?**
- **Is the mystery person *younger than* Mom?**
- **Is the mystery person *more than* 35?**
- **Is the mystery person *less than* 9?**

Experiment to find the language that is most familiar and comfortable for your students. For some, the concepts *younger* and *older* are more difficult than *more* and *less*. You may want to write some questions on the board as reminders of the preferred form, using a question mark to stand for the mystery person:

Is ? more than _____ (age)?

Is ? younger than _____ (person)?

It will probably take you a few questions to guess who the mystery person is.

Demonstrate the game by taking questions from students about the mystery person you have chosen from the grouping on the board. Reinforce the use of questions in the right form. Emphasize that when you answer each question, you must tell the truth—answering either yes or no—or the game won't work.

You have to really listen to the answers to the questions. They can help you figure out what question to ask next. You might want to use a 100 chart to help you think about the numbers.

When you are quite sure you can guess who the mystery person is, copy the name from that card on a slip of paper. When everyone in the group has written down a guess, tell who your mystery person really is. Then it's someone else's turn to pick a new mystery person.

Encourage students not to call out their guesses when they know who the mystery person is. Explain that the purpose of writing their guesses is to give everyone a chance to figure it out. Make slips of scratch paper (for the guesses) available to each group (2–3 per student). Be sure 100 charts from the previous activity are still available.

While groups are playing the game, circulate around the class to help out as needed. Plan on playing the game for about 20 minutes. Each student should have an opportunity to choose a mystery person and respond to the questions of others.

Keeping track of guesses is an issue worth raising after students have played a round or two and are more comfortable with the game.

What are some ways you are keeping track of your guesses? Is anyone using the 100 chart [or number lines]? How?

Encourage students to develop and share different strategies. For example, some might move the family portrait cards around as they get new information. Others might cross out or cover up numbers on the 100 chart or number line.

Simplifying the Game If some students are having difficulty, you can offer simplified versions of the game until they get the hang of it. For example, you can put students in pairs rather than threes so they are guessing from fewer possibilities. For an even easier game, students can mix up the family portrait cards for one group and put them facedown in a single pile. Each student takes a turn picking two cards and asking either "Who's younger?" or "Who's older?" Other group members answer the question, using any available resources (100 chart, number lines) to help them.

Observing the Students

As students are playing the game, observe how they formulate questions and whether they are listening to each other's questions and forming reasonable subsequent questions.

- How do students begin closing in on who the mystery person might be? In their guessing strategies, do they use what they know about numbers? Do they use resources, such as the 100 chart or other students, for help?
- Do they make use of their ordering of family portraits as they play the game? For example, if students learn that the mystery person is older than 15, do they group together all the people who are older than 15, eliminating the others? What happens with the next question? Do they realize that they'll still be working with people older than 15, or do they start looking through all the cards again?

To many first graders, each answer may seem like an entirely separate and unrelated clue. With experience, students will develop guessing strategies in which they use all the available information.

Activity

Organizing Data: Us and Our Siblings

This activity again uses data from the Record of Family Ages (homework, Student Sheet 8). This time, students will look at data from the entire class, focusing only on siblings.

Ask the class to help you prepare a chart on which they can put together all their data. Label a sheet of chart paper *Us and Our Siblings.* The students then help you decide what the lowest and highest age values on the chart should be.

Today we are going to make an age chart. It will show all your ages, and the ages of all your brothers and sisters. We'll start by writing numbers down one side—numbers for each age. What numbers do you think we will need? Remember, we want to show your ages *and* the ages of all of your siblings. What's the youngest age I should write? What's the oldest age I should write? (How high up should I go?)

Write the numbers that the students agree on. If some students have baby sisters or brothers who are not yet 1 year old, discuss where to put these children on your chart.

Some students may want to call them "0 years old," while others may argue for "between 0 and 1." The idea of "in between" may also come up if the students think of themselves or a sibling as being some age "and a half." Listen to everyone's ideas, giving as many students as possible a chance to say something about this before making a decision.

Distribute stick-on notes in a single color, one to each student.

Use a marker to print your name and your age on this note. Then come up and stick your note on this chart so that it shows how old you are.

This should be a very brief activity. Some students may need help locating where their note belongs on the chart.

Next, distribute stick-on notes of a different color to students who have siblings.

Make a note in this new color for each brother and sister you have. Write their name and age, just as you did on your note. Then come up and place these notes in the right place on the chart, to show how old they are.

Students who do not have siblings or who have already placed their notes may watch and talk quietly with each other about what they see developing on the chart. As needed, reinforce the stick-on note representation with transparent tape or glue to prevent data from falling off. When the chart is finished, start a whole-class discussion.

What do you notice about this data? What else can we tell from this chart?

To get this discussion going, point out certain features of the data and comment on them, and encourage students to do likewise.

There are a lot of 7's. I wonder why that is? What other numbers have a lot of names beside them? Why do those have so many?

There are a bunch of numbers where there aren't any names at all. What does that mean?

Who is the oldest sibling in our class? Who is the youngest?

Talk with students about other interesting features of the data; for example, why are notes for their siblings scattered, while theirs are in a cluster? If there are any siblings in the same age cluster as students in the class, talk about why someone the same age might not be in their class, and where else they might be. See the **Dialogue Box,** Looking at Data on Ages (p. 89), to get a sense of one group's observations about their data.

Note: Stick-on notes work well for making a rough draft, but to preserve the chart, you will later need to glue or tape each note in place.

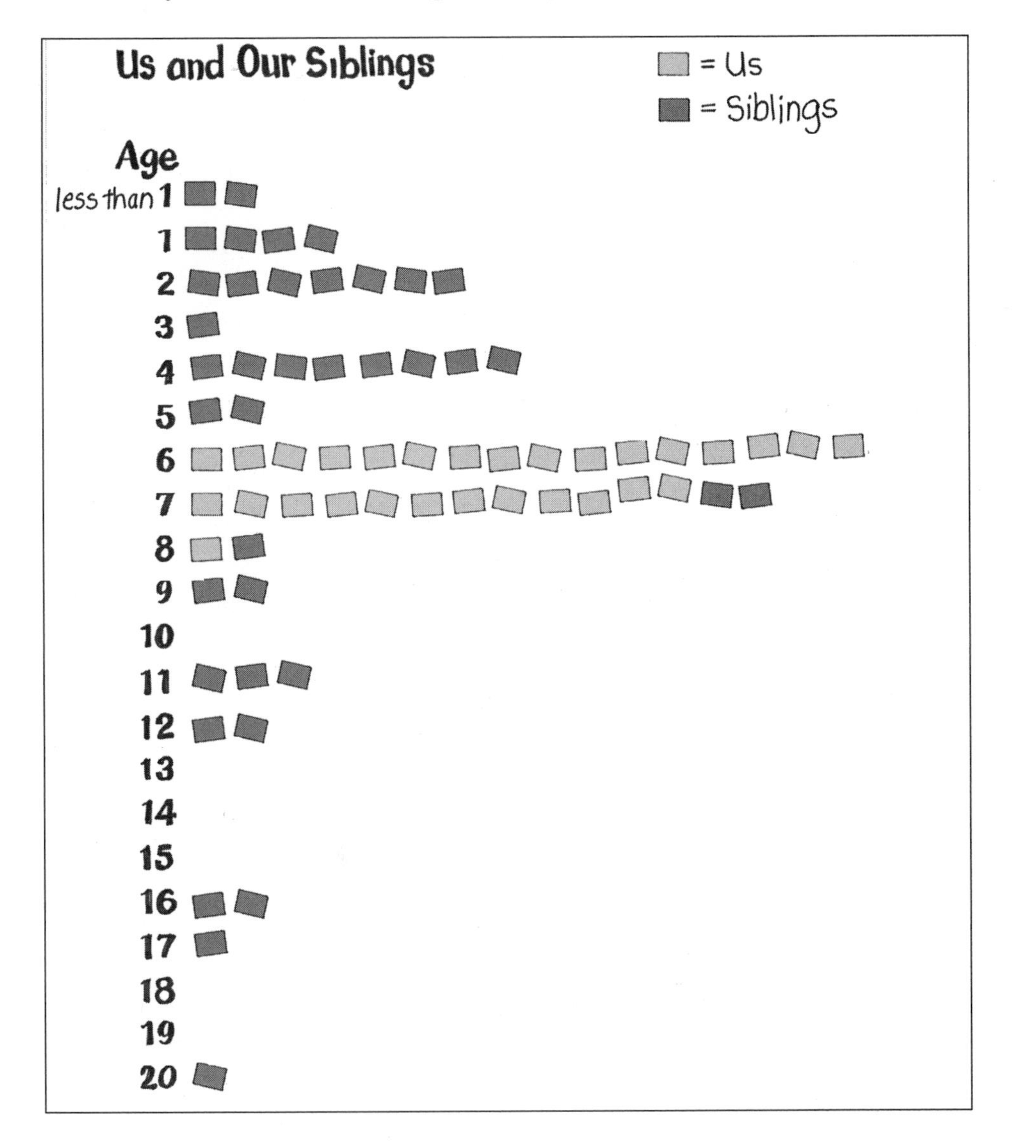

Sessions 2 and 3 Follow-Up

Family Math Problems Use the *Us and Our Siblings* chart to generate problems about number relationships. For example:

> Nadia's brother is 2. How many years will it be until he is Nadia's age?
>
> Diego has an older sister who is 11. How much older is Diego's sister than he is?

Students work in pairs with counters, cubes, 100 charts, and the class *Us and Our Siblings* chart to solve the problems.

Ordering Family Members Working in small groups, students sort their family portrait cards into sets by role; for example, moms, dads, grandparents, brothers, sisters. They then work together to put in order, by age, all the people sharing one role. Each group names something they can tell from the way they've organized their data. For example: How old is the youngest dad?

Looking at Data on Ages

These students are making observations about their class *Us and Our Siblings* chart. Aspects of particular interest to students are the longest row (the mode) and the fact that some ages do not have any people in that category. They express surprise at discovering that there are 7-year-old siblings who are not in their class, leading them to find out more about these pieces of data.

What do you notice about this chart? First of all, what is it about?

Brady: Ages.

Mia: Numbers.

Whose ages?

Claire: Boys and girls.

Luis: Your family, but not your moms and dads.

Garrett: Your brothers and sisters.

What's another word for brothers and sisters?

Nadia: Siblings.

So what does this chart about our siblings and our ages show us?

Libby: Some of them are more and some are less.

Which ones are more?

Libby: Those. The 6's.

Libby noticed that one row especially has more, the 6 row. Right? Count to yourself and see how many children are 6.

Andre: Sixteen.

Shavonne: Fifteen.

Iris: Sixteen.

[Tamika comes up and counts out loud for the class, getting 16.]

OK, 16 kids are 6 years old. That's the most on this chart. What ages had the least?

Jamaar: Number 3.

How many were there?

Jamaar: One.

Is there a number that has less than 1?

Susanna: Zero! There's nobody on 10. Ten has zero!

Do any other ages have zero?

Yukiko: Yeah, a bunch at the bottom.

So 10 and then 13, 14, 15, 18, and 19 have zero. That's the least.

Kristi Ann: I know the second highest. It's 7.

[Students count silently and discover that there are fourteen 7-year-olds.]

How many kids are 7 years old that are not in our class?

Claire: Two.

How could you tell?

Claire: Because it's blue and not yellow. On the 7.

People are really curious about who those kids are. Can anyone fill us in?

Shavonne: My twin sister is next door.

Luis: My brother is in kindergarten. He just turned 7, and I've been 7 for a long time.

Sessions 4 and 5

Attendance Comparisons

Materials

- Prepared chart for attendance data
- Rectangles of construction paper in two colors
- Glue sticks (to share)
- Class lists (available)
- Interlocking cubes (available)
- Large paper (1 sheet per student)
- Crayons or markers
- Stick-on notes, construction paper, dot stickers (optional)

What Happens

As a class, students collect attendance data. They create an "unusual day" scenario and, as an assessment, make their own representation of the attendance situation their story describes. They compare their real attendance data with their unusual day attendance data. Their work focuses on:

- describing data qualitatively ("a lot here, hardly any absent") and quantitatively ("25 here, 3 absent")
- creating a representation of the attendance on an imaginary "unusual" day
- comparing two data sets

Activity

Signing In

Pass out the paper rectangles of one color, one to each student. Have crayons or markers and glue sticks available.

Today in math class, we are going to collect attendance data using this colored paper. [Green] is for the students who are here and [orange] is for the students who are not here. All of you will write your name on the [green] paper I just gave you. I will write the name of the students who are not here on an [orange] paper.

When you have written your name, glue your paper onto this tall chart. The first person starts at the bottom, then we'll add one right above the other, the way we do when we make a stack of cubes. This way we'll make a tall tower of our attendance data, so we can see who is here and who is not.

Help students get started gluing their paper rectangles by placing the first one on the bottom of the chart paper, to the left of center. When all the students who are present have glued their rectangles in place, glue the other color (indicating those who are absent) on top of this tower.

Students spend a few minutes making observations about the data.

By looking at our attendance graph, how many students are here today? How can you tell? How many are not here? How many students are in our class?

Explore at least one way of double-checking the number of students here today. For example, you might compare the number of [green] rectangles with the number of students in class, by counting. On another sheet of chart paper, record the number of students who are here, the number who are absent, and the total.

What shall we call our graph?

Take several student suggestions before choosing one for the chart. You might call it something like *Attendance Graph,* or *Here and Not Here.* Label the tower itself and also the new chart paper with the suggestion everyone agrees on.

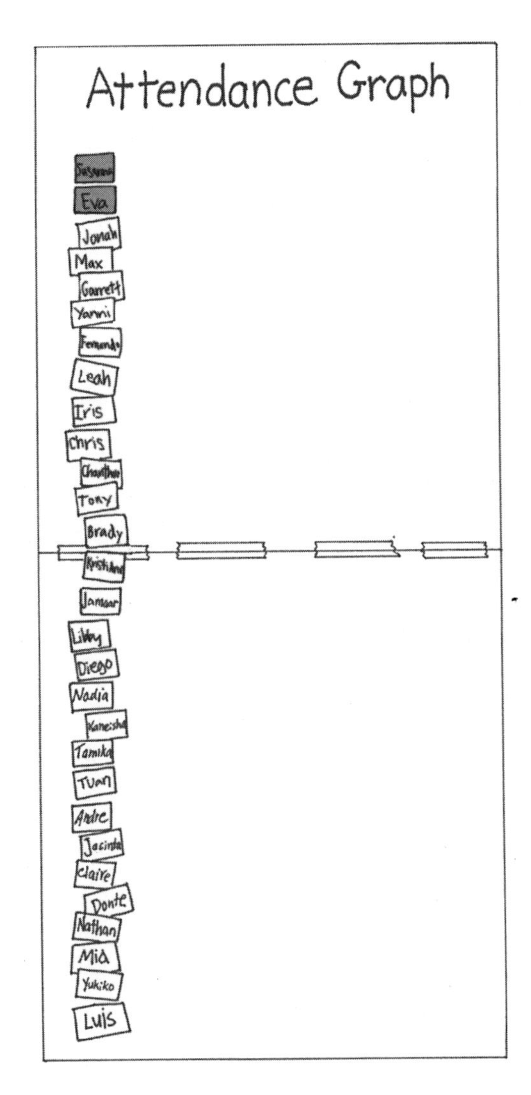

Activity

A Most Unusual Day

Usually when we collect attendance data, most of the kids are here and just a few kids are absent. What are some reasons kids might be absent?

Students briefly discuss common reasons for absence.

Today, we're going to invent a story about a very unusual day at school—a day when many of you were absent for rather strange reasons.

In this activity, students help you tell a fantasy story about who is and who is not present in their classroom on one "most unusual" day. As the story evolves, you and the students keep track of those present and absent, so that they can later make representations to show the outcomes.

As I'm telling the story, I'll pause when I get to a place where you have a chance to make up part of it. If you want to be a character in that part of the story, raise your hand and I'll call on you. Everyone will have a chance to make up part of the story about themselves if they want.

Before telling the story, label two columns on the board, *Students Who Are Here* and *Students Who Are Absent.* Explain that this is so you can keep track as you go along in the story.

Students contribute to this story by raising their hands and volunteering their own names (not other people's names). Feel free to embellish the story, personalize it, and in other ways make it exciting for your class. Adjust the numbers of students in the various situations to your class size.

❖ **Tip for the Linguistically Diverse Classroom** There are a variety of ways to aid comprehension of this story. Plan to simplify the story to its most basic elements as needed. Sketch drawings on the board to represent each "unusual situation" that keeps the students out of school (an enormous dark puddle, a tree with a giant nest, a shiny coin). Students might also use props and act out the described events as they occur.

A Most Unusual Day

Once upon a time, our class had a most unusual day. It all started one morning when I was here in the classroom, ready for you to come to school. The morning bell had rung, but the room was empty. What was going on? Could it be Saturday or Sunday? I checked the calendar. Everything seemed all right, but maybe I had overslept by a day or two. So I checked with *[name of teacher]* next door. All the students in that class were there.

Suddenly, one of you burst into the classroom. *[Ask for a student volunteer.]* ___________ looked around and asked, "What happened? Where is everyone?"

"I don't know," I said. "But until they come, you can do anything you'd like." So *[same student's name]* got busy ___________ *[ask the student to complete the sentence].*

By 9:30, a few more of you had arrived, each surprised to see the nearly empty room. The ones who came late were *[ask for volunteers] [name], [name], [name], [name],* and *[name].*

"Feel free to do anything you want until the rest of the class comes," I said as each of you came in. *[Name]* opened the closet and ___________ . *[Name]* went to the book area and ___________ . *[Name]* went down the hall and ___________ . *[Name]* and *[name]* did something they had never done in school before. They ___________ .

Well, the other children didn't show up that day at all. Here's what happened. *[Name], [name], [name],* and *[name]* were walking to school when they saw a giant puddle. It was a deep and muddy puddle, and looked a little like chocolate pudding. They peered into it and noticed something very strange. They saw ___________ . They looked and looked. And they forgot about school entirely.

Meanwhile, another group of children were coming along another street. This group included *[name], [name], [name], [name],* and *[name].* One of you, *[name a girl],* glanced up in a tree and yelled, "Look at that giant nest up there! I'm going to climb the tree and see what's in it. Slowly, *[same girl]* scooted up the tree. As she climbed, she noticed a strong smell. She held her nose, looked inside the nest, and ___________ .

"Come on up," she yelled. "You've just got to see this." So one by one, you all climbed up to the nest and spent the rest of the day ___________ .

There was one more group of you that never quite made it to school. You got to the playground, but you never made it inside. Here's what happened to *[name all the remaining students].* You were just about to come in the school door when *[name a boy]* noticed a large, shiny coin on the ground. He picked it up and started to rub it gently, only to discover that this was a magic coin. When he rubbed it, out of the coin came a ___________ . The rest of you gathered around to see, and then ___________ . That's why you never made it to school.

The ones who did come to school, *[point to all the names on the "here" list],* decided to give up and go ___________ .

And that is the story of the most unusual day our class ever had.

Activity

Assessment

Representing a Most Unusual Day

At the end of the story, turn to your *Here* and *Absent* lists on the board and ask if everyone in the class is accounted for. If anyone has been left out (for example, a child who is really absent or out of the room), add a postscript to include that person in the story.

These lists show our attendance data on "a most unusual day." Each of you is going to find your own way to represent this attendance data.

How many were actually in school on our most unusual day? How many were absent? How many students will you include in your representation?

As students respond, ask them to describe their strategies for figuring this out. Some may say they counted the list of names on the board. Others may use what they already know about the number of students in their class. For example, "I know that we usually have 26 kids, so there must be 26 names."

How do you think you could keep track of whether you've got everyone represented?

Discuss ways of keeping track, most of which will involve using the list on the board. Distribute large paper, crayons or markers, and any other materials you are supplying for creating representations, such as stick-on notes, sticky dots, or construction paper scraps. Make class lists available to anyone who wants one.

Observing the Students

Watch as students are working on their representations.

- How do students keep track of the number of children in each group? Do they count accurately ?
- What ideas are they using to make their representations clear to others? Does their representation communicate that this is attendance data? that it is *unusual* attendance data?
- How do they show the number of students present and absent? Are their categories distinct? Can someone tell that one group is bigger than another without counting?

Students work on their representations through Session 4 and into the beginning of Session 5.

As students finish, they pair up to show each other their work. Ask them to each name one thing that is very clear about their partner's representation and one thing that needs further clarification. Students then revise their work.

See the **Teacher Note,** Assessment: Representing a Most Unusual Day (p. 98) for examples of students' representations.

Activity

Comparing Usual and Unusual Days

Gather students where everyone can see the attendance chart the class made in Session 4. They should have with them their representations of attendance on the unusual day.

Can someone remind us what this chart we made yesterday shows? Would you say this shows a fairly usual day at school?

We're going to compare attendance on our usual day and our most unusual day. To help us, we are going to make another tower on this chart paper using the same kind of colored paper. Our new tower will show the attendance in our class on our most unusual day.

Think about the story we made up together. If you were *here* on the unusual day, what color paper will you need? What color paper if you were *not here* on the unusual day?

Distribute papers of the appropriate color to each student. Students may use their representations or your lists (if still on the board) to help them remember if they were "here" or not. Invite the students who were "here" to glue their rectangles onto the chart paper, starting from the bottom and going up. If there are any students absent, you will need to place paper of the appropriate color for them. Next invite the students who were "absent" to glue their rectangles above the others, making the tower taller. Label this new tower something like "Our Unusual Day."

On our unusual day, how many students were here? How many were gone?

Ask for a few different strategies to double-check this information. Students may count the rectangles, look at their own representations, or simply remember this data. Record these numbers on the chart you began in Session 4, under the headings *Here, Absent,* and *Total.*

How were these two days of attendance different?

Students may speak of graphical features, perhaps noticing a lot of [green] names for the real day and only a few for the unusual day. Some students may focus on specifics, saying things like "Three students were absent on the real day, and 15 were absent on the unusual day."

What's the same on both days?

Students may notice that the two colors are the same, or that each tower has the same number of rectangles (even though the towers may be slightly different heights).

Now let's look at some of your representations. Did anyone create a representation that is like this tower? If you did, hold up your work. How is yours like this tower?

Let's look at some that seem different from these towers. What can you tell from these that you can't tell from the towers?

As students respond, be sure everyone else can see the representation being talked about. Discuss with the students what seems especially clear or unique about these representations.

What If? After a few students have shared their work, focus attention on the numbers and the relationship between parts and wholes by asking a series of "What if?" questions. Students use their attendance representations and cubes to count and figure out solutions to the "What if?" questions. You might repeat the scenarios from the unusual day story, or you might make up new ones. For each "What if?" record on chart paper the numbers for *here, absent,* and *total.*

- What if 16 of you went to the kitchen to decorate a birthday cake for the principal and the rest of you stayed here in class to practice singing a birthday song. How many would be here singing?
- What if 13 of you stayed here to build a spacecraft, and the rest of you went on the Magic School Bus to the center of the earth. How many of you would be gone?
- What if 28 of you went to see a play of *The Wizard of Oz.* How many would be left here in class?
- What would change on this graph if a new student joins our class?

Students might want to contribute their own "What if?" questions and imaginary situations.

❖ **Tip for the Linguistically Diverse Classroom** Make simple sketches to represent each "What if?" situation to aid in comprehension.

Activity

Choosing Student Work to Save

As the unit ends, you may want to use one of the following options for creating a record of students' work on this unit.

- Students look back through their collection of work and think about what they learned in this unit, what they remember most, what was hard or easy for them. You might have students discuss this with a partner or have students share in the whole group.
- Depending on how you organize and collect student work, you may want to have students select some favorite pieces to keep in their portfolio. In addition, you may want to choose some examples from each student's math folder to include. Some useful pieces for assessing student growth over the school year might include the Representations of Sorted Shapes (p. 9), posters of A Favorite Sorting Rule (p. 29), Representations of Eagle and Whale Data (p. 36), students' plans, representations, and findings for the survey data they collected during Investigation 2 (p. 48), and their work on Representing a Most Unusual Day (p. 94).
- If you send a selection of work home for families to see, you might include a short letter summarizing the work in this unit. You could enlist the help of your students and together generate a letter that describes the mathematics that their work shows. If you are keeping a year-long portfolio of mathematics work for each student, this work should be returned to you. Alternatively, you might keep the originals and make copies for students to take home.

Teacher Note

Assessment: Representing a Most Unusual Day

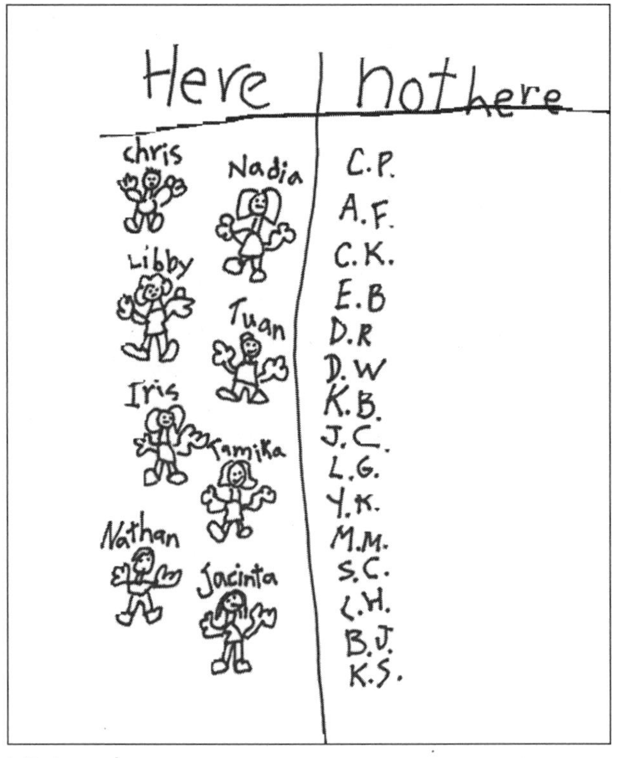

Mia's work

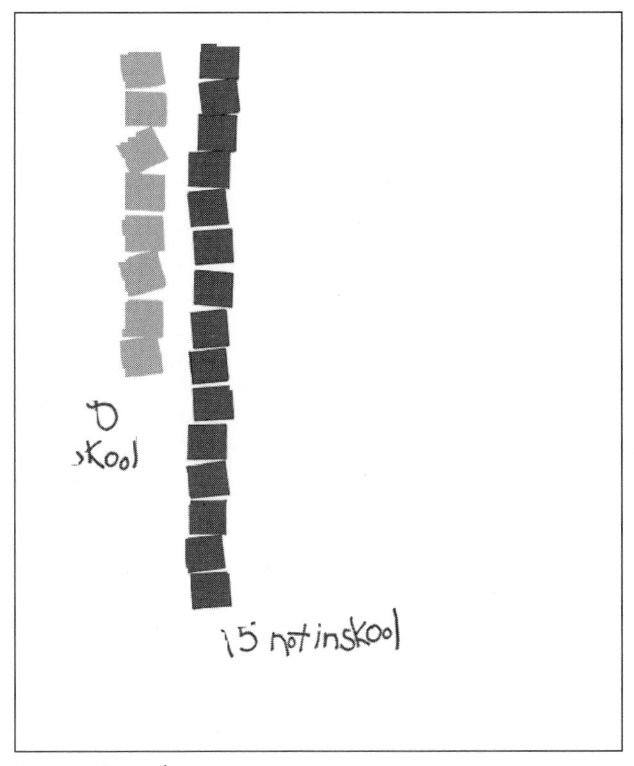

Jamaar's work

Diego's work

After creating a class set of attendance data for "a most unusual day," students find their own ways to represent this data. As you review student work for this assessment activity, look for evidence of the important ideas about representing data that students have been exploring:

- Are students' representations clear to others?
- Does the work clearly communicate what the data tell about? Do students use pictures? words? both?
- Are their categories visually distinctive?
- Is it easy to compare the sizes of the categories? How have students used numbers, words, or pictures to clarify this information?

Mia, Jamaar, and Diego have created different representations, but each has clear, distinctive categories and communicates the data quite well. Each of these representations is accurate, and the students use words to tell what each group represents. Mia has preserved the names (or initials) of the students in each group, Diego includes the total number of students, and

Teacher Note

continued

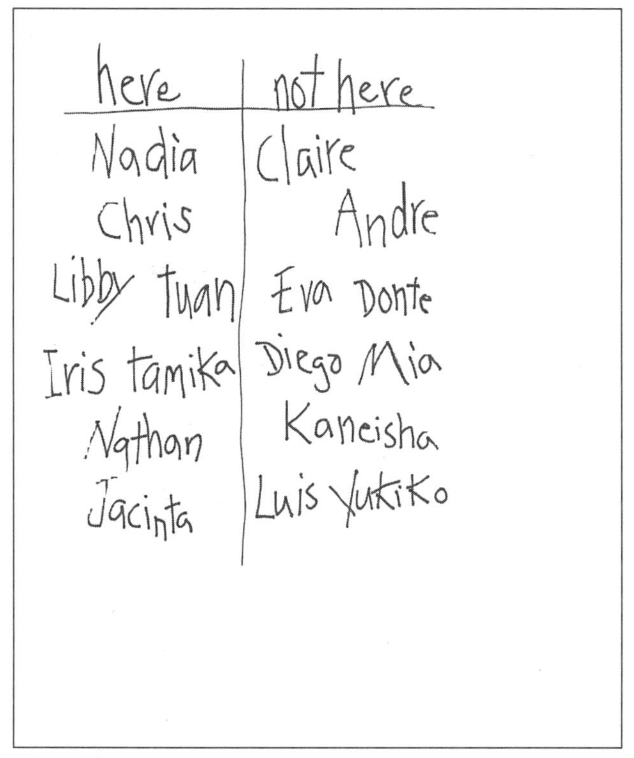

Andre's work

Libby's work

Jamaar uses columns of squares as well as numbers to show the relative sizes of the categories.

Some students may need additional support as they are working. For example, you may find students who simply copy the representation (the here/not here list) that is on the board, as Andre was starting to do.

The teacher asked Andre to think of another way to represent the data. She reminded him what was available (stick-on notes, construction paper) and suggested he draw pictures or chose another material to show how many students were here and not here on the unusual day.

Some students may not organize the data so that someone else can clearly see how many groups there are and which pieces of data belong in each group.

When the teacher asked Libby which pieces of data belonged together, she could readily point to the stick-on notes in each group. The teacher pointed out that someone who hadn't heard the unusual day story wouldn't be able to tell from Libby's representation who was here and who was not. Libby was asked to find a way to clearly show others which students belong in which group. She then revised her work, adding labels and boundaries so it became clear which students were in the *Here* and *Not Here* groups.

As you think back over students' work in this unit, also consider the following:

- How have students' representations changed over the course of the unit? Are they more comfortable constructing their own data representations? trying out new approaches?
- What strategies are students using to keep track of their data? Are their strategies working? Are they generally accurate in their counts?
- Are students able to use their representation to explain or present the results to you? to classmates?
- Are students able to interpret some of their classmates' representations?

Counting

Counting is an important focus in the grade 1 *Investigations* curriculum, as it provides the basis for much of mathematical understanding. As students count, they are learning how our number system is constructed, and they are building the knowledge they need to begin to solve numerical problems. They are also developing critical understandings about how numbers are related to each other and how the counting sequence is related to the quantities they are counting.

Counting routines can be used to support and extend the counting work that students do in the *Investigations* curriculum. As students work with counting routines, they gain regular practice with counting in familiar classroom contexts, as they use counting to describe the quantities in their environment and to solve problems based on situations that arise throughout the school day.

How Many Are Here Today?

Since you must take attendance every day, this is a good time to look at the number of students in the classroom in a variety of ways.

Ask students to look around and make an estimate of how many are here today. Then ask them to count.

At the beginning of the year, students will probably find the number at school today by counting each student present. To help them think about ways to count accurately, you can ask questions like these:

How do we know we counted accurately? What are different methods we could use to keep track and make sure we have an exact count? (For example, you could count around a circle of seated students, with each student in turn saying the next number. Or, all students could start by standing up, then sit down in turn as each says the next number.)

Is there another way we could count to double-check? (For example, if you counted around the circle one way, you could count around the circle the other way. If you are using the standing up/sitting down method, you could recount in a different order.)

You might want to count at other times of the day, too, especially when several students are out of the room. For example, suppose groups of students are called to the nurse's office for hearing examinations. Each time a new group of students leaves, you might ask the class to look around and think about how many students are in the room now:

So, this time Diego's table and Mia's table both went to the nurse. Usually we have 28 students here. Look around. What do you think? Don't count. Just tell me about how many students might be here now. Do you think there are more than 5? more than 10? more than 20?

Later in the year, some students may be able to use some of the information they know about the total number of students in the class and how many students are absent to reason about the number present. For example, suppose 26 students are in class on Monday, with 2 students absent. On Tuesday, one of those students comes back to school. How many students are in class today? Some students may still not be sure without counting from one, but other students may be able to reason by counting on or counting back, comparing yesterday and today. For example, a student might solve the problem in this way:

> "Yesterday we had 26 students, and Michelle and Chris were both absent. Today, Chris came back, so we have one more person, so there must be 27 today."

Another might solve it this way:

> "Well we have 28 students in our class when everyone's here. Now only Michelle is absent, so it's one less. So it's 27."

From time to time, you might keep a chart of attendance over a week or so, as shown below. This helps students become familiar with different combinations of numbers that make the same total. If you have been doing any graphing, you might want to present the information in graph form.

Day	Date	Present	Absent	Total
Monday	March 2	26	2	28
Tuesday	March 3	27	1	28
Wednesday	March 4	27	1	28
Thursday	March 5	27	1	28
Friday	March 6	28	0	28
Monday	March 9	28	0	28
Tuesday	March 10	26	2	28
Wednesday	March 11	25	3	28

After a week or two, look back over the data you have collected. Ask questions about how things have changed over time.

In two weeks of attendance data, what changes? What stays the same?

On which day were the most students here? How can you tell? Which day shows the least students here? What part of the [chart] gives you that information?

Another idea (for work with smaller numbers) is to keep track of the number of girls and boys present and absent each day. Again, many students will count by 1's. Later in the year, some will also reason about these numbers:

> "There are two people absent today and they're both girls. We usually have 14 girls, and Kaneisha's sick, that's 13, and Claire's sick, that's 12."

Can Everyone Have a Partner?

Attendance can be an occasion for students to think about making groups of two:

We have 26 students here today. Do you think that everyone can have a partner if we have 26 students?

Students can come up with different strategies for solving this problem. They might draw 26 stick figures, then circle them in 2's. They could count out 26 cubes, then put them together in pairs. They might arrange themselves in 2's, or count by 2's .

At the beginning of the year, many of your students will need to count by 1's from the beginning each time you add two more students, but gradually some will begin to notice which numbers can be broken up into pairs:

> "I know 13 doesn't work, because you can do it with 12, and 13's one more, so you can't do it."

Some students will begin to count by 2's, at least for the beginning of the counting sequence. Then, as the numbers get higher, they may still be able to keep track of the 2's, but need to count by 1's:

> "So, that's 2, 4, 6, 8, 10, 12, um, 13, 14 . . . 15, 16."

As you explore 2's with your students, keep in mind that many of them will need to return to 1's as a way to be sure. Even though some students learn the counting sequence 2, 4, 6, 8, 10, 12 . . . by rote, they may not connect this counting sequence to the quantities it represents at each step.

One teacher found a way to help students develop meaning for counting by 2's. She took photographs of each student, backed them with cardboard, then used them during the morning meeting as a model for making pairs. She laid out the photos in two columns, and asked about the new total after the addition of each pair:

We have 10 photos out so far. The next two photos are for William and Yanni. When we put those two photos down, how many photos will we have?

Lining up is another time to explore making pairs. Before lining up, count how many students are in class (especially if it's different from when you took attendance). Ask students whether they think the class will be able to line up in even pairs. For many first grade students, the whole class is too many people to think about. You can ask about smaller groups:

What if Kristi Ann's table lines up first? Do you think we could make even partners with the people at that table?

What about Shavonne's table? ... Do you think Shavonne's table will have an extra? How do you know?

Is there another table that would have an extra that we could match up with the extra person from Shavonne's table?

Once students are lined up in pairs, they can count off by 2's. Because most first graders will need to hear all the numbers to keep track of how the counting matches the number of people, ask them to say the first number in the pair softly and the second one loudly. Thus the first pair in line can say, "1, **2**," the second pair can say, "3, **4**," and so forth.

Counting to Solve Problems

Be alert to classroom activities that lend themselves to a regular focus on solving problems through counting. Use these situations as contexts for counting and keeping track, estimating small quantities, breaking quantities into parts, and solving problems by counting up or back. For example, take a daily milk count:

Everyone who is buying milk today stand up. Without counting yet, who has an idea how many students might be standing up? Is it more than 5? more than 10? more than 50? ... Now, let's count. How could we keep track today so that we get an accurate count?

You can make a problem out of lunch count:

We found out that 23 students are buying school lunch today. We have 27 students here. So how many students brought their own lunch from home today?

Watch for the occasional sharing situation:

Claire brought in some cookies she made to share for snack. She brought 36 cookies. Is that enough for everyone to have one cookie, including me and our student teacher? Oh, and Claire wants to invite her little brother to snack. Do we have enough for him, too? Will there be any cookies left over?

The sharing of curriculum materials can also be the basis of a problem:

Each pair of students needs a deck of number cards to share. While I'm getting things together, work on this problem with your partner. We said this morning that we have 26 students here. If I need one deck for each pair, how many decks do I need?

Exploring Data

Through data routines at grade 1, students gain experience working with categorical data—information that falls into categories based on a common feature (for example, a color, a shape, or a shared function). The data routines specifically extend work students do in the *Investigations* curriculum. The Guess My Rule game and its many variations (introduced in the unit *Survey Questions and Secret Rules*) can be used throughout the year for practice with organizing sets into categories and finding ways to describe those categories—a fundamental part of analyzing data. Students can also practice collecting and organizing categorical data with quick class surveys that focus on their everyday experiences; this practice supports the survey-taking they do in the curriculum.

Guess My Rule

Guess My Rule is a classification game in which players try to figure out the common characteristic, or attribute, of a set of objects. To play the game, the rule maker (who may be the teacher, a student, or a small group) decides on a secret rule for classifying a particular group of things. For example, a rule for classifying people might be WEARING STRIPES.

The rule maker (always the teacher when the game is first being introduced) starts the game by giving some examples of objects or people who fit the rule. The guessers then try to find other items that fit the same rule. Each item (or person) guessed is added to one of two groups—either *does fit* or *does not fit* the rule. Both groups must remain clearly visible to the guessers so they can make use of all the evidence as they try to figure out the rule.

Emphasize to the players that "wrong" guesses are as important as "right" guesses because they provide useful clues for finding the rule. When you think most students know the rule, ask for volunteers to share their ideas with the class.

Once your class is comfortable with the activity, students can choose the rules. Initially, you may need to help students choose appropriate rules.

Guess my Rule with People When sorting people according to a secret rule, always base the rule on just one feature that is clearly visible, such as WEARING A SHIRT WITH BUTTONS, or WEARING BLUE. When students are choosing the rule, they may choose rules that are too obvious (such as BOY/GIRL), so vague as to apply to nearly everyone (WEARING DIFFERENT COLORS), or too obscure (HAS AN UNTIED SHOELACE). Guide and support students in choosing rules that work.

Guess My Rule with Objects Class sets of attribute blocks (blocks with particular variations in size, shape, color, and thickness) are a natural choice for Guess My Rule. You can also use collections of objects, such as sets of keys, household container lids, or buttons. One student sorts four to eight objects according to a secret rule. Others take turns choosing an object from the collection that they think fits the rule and placing it in the appropriate group. If the object does not fit, the rule maker moves it to the NOT group. After several objects have been correctly placed, students can begin guessing the rule.

Guess My Object Once students are familiar with Guess My Rule, they can use the categories they have been identifying to play another guessing game that also involves thinking about attributes. In this routine, students guess, by the process of elimination, which particular one of a set of objects has been secretly chosen. This works well with attribute blocks or object collections.

To start, place about 20 objects where everyone can see them. The chooser secretly selects one of the objects on display, but does not tell which one (you may want the chooser to tell you, privately). Other students ask yes-or-no questions, based on attributes, to get clues to help them identify the chosen object. After each answer, students move to one side the objects that have been eliminated. That is, if someone asks "Is it round?" and the answer is yes, all objects that are *not* round are moved aside.

Pause periodically to discuss which questions help eliminate the most objects. For example, "Is it this one?" eliminates only one object, whereas "Is it red?" may eliminate several objects. For more challenge, students can play with the goal of identifying the secret object with the fewest questions.

Quick Surveys

Class surveys can be particularly engaging when they connect to activities that arise as a regular part of the school day, and they can be used to help with class decisions. As students take surveys and analyze the results, they get good practice with collecting, representing, and interpreting categorical data.

Early in first grade, to keep the surveys quick and the routine short, use questions that have exactly two possible responses. For example:

Would you rather go outside or stay inside for recess today?

Will you drink milk with your lunch today?

Do you need left-handed or right-handed scissors?

As the school year progresses, you might include some survey questions that are likely to have more than two responses:

Which of these three books do you want me to read for story time?

Who was your teacher last year?

Which is your favorite vegetable growing in our class garden?

How old are you?

In which season were you born?

Try to choose questions with a predictable list of just a few responses. A question like "What is your favorite ice cream flavor?" may bring up such a wide range of responses that the resulting data is hard to organize and analyze.

As students become more familiar with classroom surveys, invite the class to brainstorm questions with you. You may decide to avoid survey questions about sensitive issues such as families, the body, or abilities, or you might decide to use surveys as a way of carefully raising some of these issues. In either case, it is best to avoid questions about material possessions ("Does your family have a car?").

Once the question is chosen, decide how to collect and represent data. Be sure to vary the approach. One time, you might collect data by recording students' responses on a class list. Another time, you might take a red interlocking cube for each student who makes one response, a blue cube for each student who makes the other response. Another time, you might draw pictures. If you have prepared Kid Pins and survey boards for use in *Mathematical Thinking at Grade 1,* these can be used for collecting the data from quick surveys all year.

Initially, you may need to help students organize the collected data, perhaps by stacking cubes into "bars" for a "graph," or by making a tally. Over time, students can take on more responsibility for collecting and organizing the data.

Always spend a little time asking students to describe, compare, and interpret the data.

What do you notice about these data?

Which group has the most? the least? How many more students want [recess indoors today]?

Why do you suppose more would rather [stay inside]? Do you think we'd get similar data if we collected on a different day? What if we did the same survey in another class?

Understanding Time and Changes

These routines help students develop an understanding of time-related ideas such as sequencing of events, understanding relationships among time periods, and identifying important times in their day.

Young students' understanding of time is often limited to their own direct experiences with how important events in time are related to each other. For example, explaining that an event will occur *after* a child's birthday or *before* an important holiday will help place that event in time for a child. Similarly, on a daily basis, it helps to relate an event to a benchmark time, such as *before* or *after* lunch. Both calendars and daily schedules are useful tools in sequencing events over time and preparing students for upcoming events. These routines help young students gain a sense of basic units of time and the passage of time.

Calendar

The calendar is a real-world tool that people use to keep track of time. As students work with the calendar, they become more familiar with the sequence of days, weeks, and months, and the relationships among these periods of time. Calendar activities can also help students become more familiar with relationships among the numbers 1–31.

Exploring the Monthly Calendar At the start of each month, post the monthly calendar and ask students what they notice about it. Some students might focus on arrangement of numbers or total number of days, while others might note special events marked on the calendar, or pictures or designs on the calendar. All these kinds of observations help students become familiar with time and ways that we keep track of time. You might record students' observations and post them near the calendar.

As the year progresses, encourage students to make comparisons between the months. Post the calendar for the new month next to the calendar for the month just ending and ask students to share their ideas about how the two calendars are similar and different.

Months and Years To help students see that months are part of a larger whole, display the entire calendar year on a large sheet of paper. Cut a small calendar into individual monthly pages and post the sequence of months on the wall. You might decide to post the months according to the school year, September through August, or the calendar year January through December. At the start of each month, ask students to find the position of the new month on the larger display. From time to time, you might also use this display to point out dates and distances between them as you discuss future events or as you discuss time periods that span a month or more. (Last week was February vacation. How many weeks until the next vacation?)

How Many More Days? Ask students to figure out how long until special events, such as birthdays, vacations, class trips, holidays, or future dates later in the month. For example:

Today is October 5. How many more days until October 15?

How many more days until [Nathan's] birthday?

How many more days until the end of the month?

Ask students to share their strategies for finding the number of days. Initially, many students will count each subsequent day. Later, some students may begin to find their answers by using their growing knowledge of calendar structure and number relationships:

> "I knew there were three more days in this row and I added them to the three days in the next row. That's 6 more days."

Others may begin using familiar numbers such as 5 or 10 in their counting:

> "Today is the 5th. Five more days is 10, and five more is 15. That's 10 more days until October 15."

For more challenge, ask for predictions that span two calendar months. For example, you might post the calendar for next month along side of the calendar of this month and ask a question like this:

It's April 29 today. How many more days until our class trip on May 6?

Note that we can refer to a date either as October 15 or as the 15th day of October. Vary the way you refer to dates so that students become comfortable with both forms. Saying "the 15th day of October" reinforces the idea that the calendar is a way to keep track of days in a month.

How Many Days Have Passed? Ask questions that focus on events that have already occurred:

How many days have passed since [a special event]? since the weekend? since vacation?

Mixed-Up Dates If your monthly display calendar has date cards that can be removed or rearranged, choose two or three dates and change their position on the calendar so that the numbers are out of order. Ask students to fix the calendar by pointing out which dates are out of order.

Groups of two or three can play this game with each other during free time. Students can also remove all the date cards, mix them up, and reassemble the calendar in the correct order. You might mark the space for the first day of the month so that students know where to begin.

Daily Schedule

The daily schedule narrows the focus of time to hours and shows students the order of familiar events over time. Working with schedules can be challenging for many first graders, but regular opportunities to think and talk about the idea will help them begin predicting what comes next in the schedule. They will also start to see relationships between particular events in the schedule and the day as a whole.

The School Day Post a schedule for each school day. Identify important events (start of school, math, music, recess, reading, lunch) using pictures or symbols and times. Include both analog (clock face) and digital (10:15) representations. Discuss the daily schedule each day with students using words such as *before* math, *after* recess, *during* the morning, *at the end of* the school day. Later in the school year you can begin to identify the times that events occur as a way of bridging the general idea of sequential events and the actual time of day.

The Weekend Day Students can create a daily schedule, similar to the class schedule, for their weekend days. Initially they might make a "timeline" of their day, putting events in sequential order. Later in the year they might make another schedule where they indicate the approximate time of day that events occur.

Weather

Keeping track of the weather engages young students in a real-life data collection experience in which the data they collect changes over time. By displaying this ongoing collection of data in one growing representation, students can compare changes in weather across days, weeks, and months, and observe trends in weather patterns, many of which correspond to the seasons of the year.

Monthly Weather Data With the students, choose a number of weather categories (which will depend on your climate); they might include sunny, cloudy, partly cloudy, rainy, windy, and snowy.

If you vary the type of representation you use to collect monthly data, students get a chance to see how similar information can be communicated in different ways. On the following page you'll see some ways of representing data that first grade teachers have used.

At the end of each month (and periodically throughout the month), ask questions to help students analyze the data they are collecting.

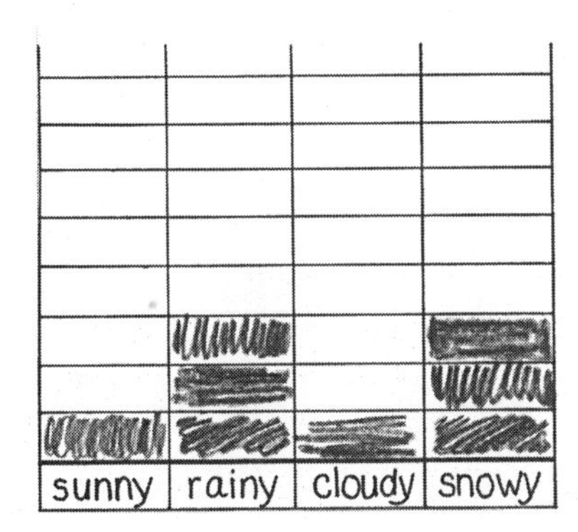

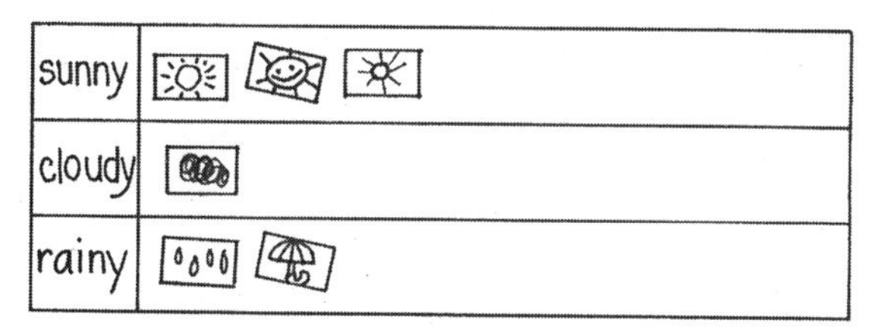

sun	clouds
rain	snow

Weather data can be collected on displays like these. In the second example, a student draws each day's weather on an index card to add to the graph. The third example uses stick-on dots.

What is this graph about?

What does this graph tell us about the weather this month (so far)?

What type of weather did we have for the most days? What type of weather did we hardly ever have?

How is the weather this month different from the weather last month? What are you looking at on the graph to help you figure that out?

How do you think the weather graph for next month will look?

Yearly Weather Data If you collect and analyze weather data for some period of time, consider extending this over the entire school year. Save your monthly weather graphs, and periodically look back to see and discuss the changes over longer periods of time.

Another approach over the entire year is to prepare 10-by-10 grids from 1-inch graph paper, making one grid for each weather category your class has chosen. Post the grids, labeled with the identifying weather word. Each day, a student records the weather by marking off one square on one or more grids; that is, on a sunny day, the student marks a square on the "sunny" grid, and if it's also windy, he or she marks the "windy" grid, too.

From time to time, students can calculate the total number of days in a certain category by counting the squares. Because these are arranged in a 10-by-10 grid, some students may use the rows of 10 to help them calculate the total number of days. ("That's 10, and another 10 is 20, and 21, 22, 23.")

Making Weather Representations After students have had some experience collecting and recording data in the grade 1 curriculum (especially in *Survey Questions and Secret Rules*), they can make their own representation of the weather data. For one month, record the weather data on a piece of chart paper (or directly on your monthly calendar), without organizing it by category. At the end of the month, ask students to total the number of sunny days, rainy days, and so forth, and post this information (perhaps as a tally). Students then make their own representation of the data, using pictures, numbers, words, or a combination of these. Encourage them to use clear categories and show the number of days in each.

The following activities will help ensure that this unit is comprehensible to students who are acquiring English as a second language. The suggested approach is based on *The Natural Approach: Language Acquisition in the Classroom* by Stephen D. Krashen and Tracy D. Terrell (Alemany Press, 1983). The intent is for second-language learners to acquire new vocabulary in an active, meaningful context.

Note that *acquiring* a word is different from *learning* a word. Depending on their level of proficiency, students may be able to comprehend a word upon hearing it during an investigation, without being able to say it. Other students may be able to use the word orally, but not read or write it. The goal is to help students naturally acquire targeted vocabulary at their present level of proficiency.

We suggest using these activities just before the related investigations. The activities can also be led by English-proficient students.

Investigation 1

set, group, go together, similar, alike

1. Show the group a mixed collection of two types of common classroom items, such as chalk and pencils. Explain that you and the students are going to *group* these things into two *sets.*
2. Put 2–3 pieces of chalk together and 2–3 pencils together.

   These go together. This set is chalk. These go together. This set is pencils.
3. Explain that the pieces of chalk are *similar,* and the pencils are similar. Ask students to find more objects in your mixed collection that *go together* with one set or the other.
4. Continue by asking questions about the two sets.

   Which group does this *[hold up a pencil]* **go in?**

   How are the things in this set *[point to the pencils]* **similar?**

   How are the things in the other set alike?
5. Next, hold up a crayon. Ask if this would go together with either set, or if you should start a new group. If they indicate the crayon should go in one of the established sets, ask why. If they want to start a new group, ask why they think the crayon does not go with either set.

Investigation 3

month (and month names), order

1. Randomly show pages from a collection of different calendars. Read aloud all the month names (in no particular order).
2. Ask students if any month name is familiar to them, and why.
3. Show a small calendar of all twelve months together. Then offer an identical or similar calendar with the months cut apart in little squares. Mix them up, and explain that the cut-up months are now out of *order.*
4. Ask students to work together to put the months in the same order shown in the whole calendar. Model ordering the first two or three months if needed. (Note that this is intended as a simple matching activity, with no expectation that students will recognize the order of the months.)

most, least

1. Ask three students to stand before the group. Give the first *five* books to hold; give another *three* books to hold, and give the last student *one* book.
2. Tell the group that [name of student] has the *most* books, and that [name of student] has the *least* books.
3. Continue the activity by replacing the books with other objects that you have enough of (such as rulers or playground balls). Each time, ask:

   Who has the most? Who has the least?
4. Choose another object, such as pencils, and ask a volunteer to distribute them to the three students. Say which student should get the *most* and which should get the *least.* Ask the rest of the group to verify whether the objects were correctly distributed.

Blackline Masters

______________________, 19____

Dear Family,

As part of our mathematics program this year, we will be working on a math unit about *data*—the facts, or the information, with which we differentiate and describe people and things in our world. At this level, the children will be sorting, collecting, and making representations as they learn to "tell the story of the data."

For the sorting activities, the children will be sorting attribute blocks (by shape, by color, and by size), and collections of household things, like buttons and lids. Defining categories for sorting is a key skill in both math and science.

After working with collections for a while, the children will practice gathering data through survey questions they ask each other—things like "Would you rather eat ice cream in a cone or a cup?" and "Do you button your shirt starting at the top or the bottom?" They'll also be exploring data about the different birth dates in our class, and the ages of people in their families.

While we are working with data, the children will continue to develop number skills. They practice counting quantities and determining which are *more* and which are *less*.

There are many ways you can extend this unit at home. Sorting comes up every day, and young children are often eager to help sort laundry, items for recycling, groceries, hardware, and spare change. Encourage your child to try different ways of sorting. Explain how your own sorting systems work and why you use these. Also, if your child collects something, you might work together to organize or sort the collection.

You can also work with data as numbers. When children are gathering data about ages of people in their families, talk with them about who's older than they are, and by how much. If there are siblings, you could ask questions about the combined ages of the children. In either case, encourage your child to figure out different strategies for comparing or adding up.

Above all, have fun with this unit at home as you watch your children learn.

Sincerely,

Name Date

Student Sheet 1

Describe a Button

1. Find a button. Glue, tape, or draw it below.
2. What are some words that describe your button? List as many as you can.

My button

Name Date

Student Sheet 2

Sorting Things

In class, we are sorting things. Here is an example of a way to sort shapes:

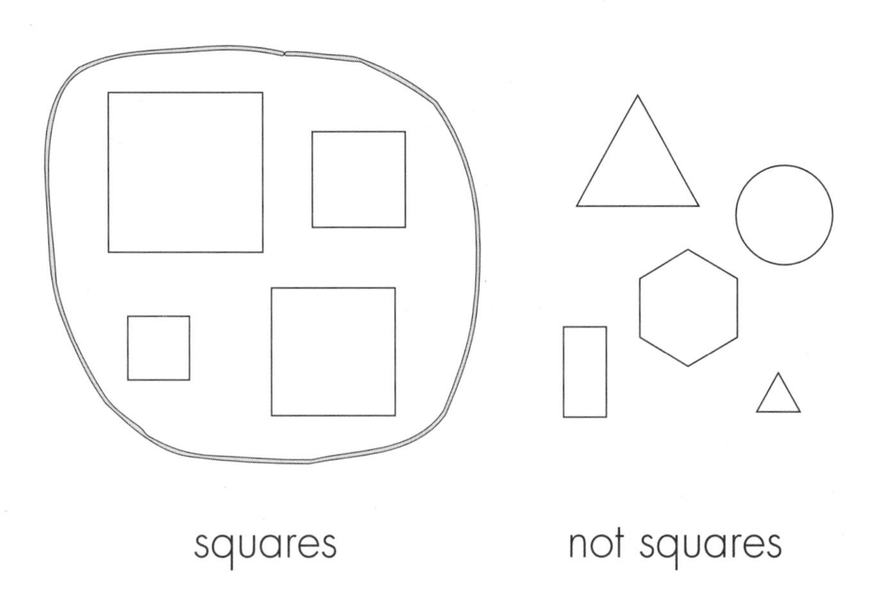

squares not squares

1. Find a group of 10 similar things at home. Here are some ideas: buttons, coins, stamps, things to write with, things to recycle.
2. What did you find to sort?

____

3. Sort your things into two groups. Try a few different ways to group them. Then choose the way you like best.

   On the back of this sheet, draw a picture that shows how you sorted your things.
4. Write words that tell how you sorted them.

Name Date

Student Sheet 3

Design a Lid

Draw a lid. Draw it very large. You can make up a lid or copy a real lid.

Think about our class lid collection. How is your lid like those lids? List two or three groups your lid would fit in.

____

____

____

____

ATTRIBUTE SHAPE CARDS

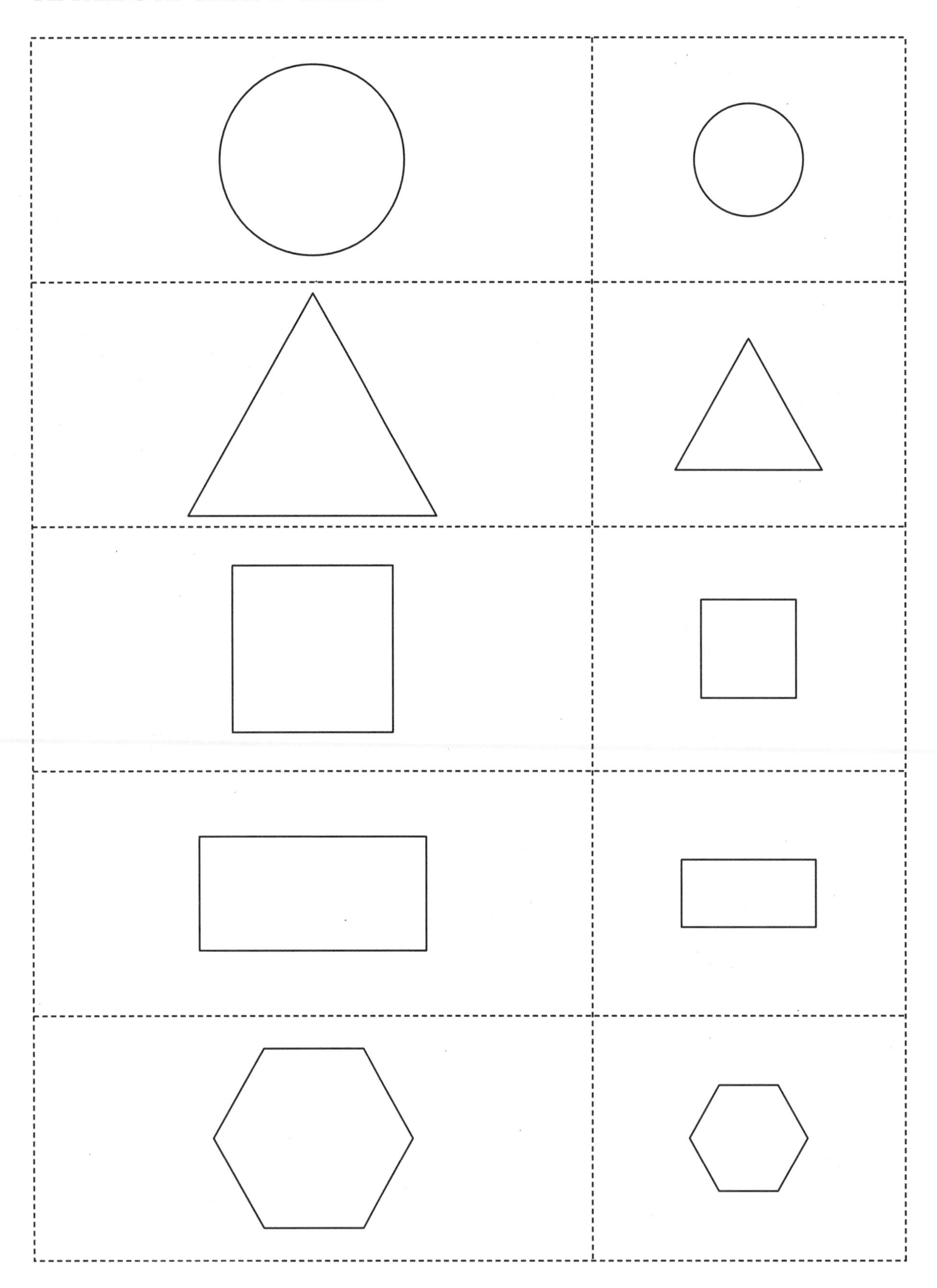

Name Date

Student Sheet 4

Our Plan for Collecting Data

1. What is your question?

____

____

2. Who will ask the question?

____

3. Who will record the answers?

____

4. How will you record the answers?

____

____

____

____

5. How will you make sure you asked everyone?

____

____

____

____

Name Date

Student Sheet 5

Collections for Sorting

Find something at home you can bring to class for sorting activities. You collection should have 10 to 15 things that are similar, but not identical. Here are some ideas:

stickers
rocks or pebbles
sports cards
marbles
used envelopes
fabric scraps
shells
stamps
old keys
leaves
seeds
pencils

Name Date

Student Sheet 6

Our Findings

1. What was your question?

2. What did you find out?

3. What surprised or interested you?

SURVEY QUESTIONS (page 1 of 2)

Would you rather… be invisible? or be able to fly?	Would you rather… be able to talk to animals? or live forever?
Would you rather… have the power to see through things? or the power to move things with your eyes?	Would you rather… have super strength? or have a super brain?
Would you rather… visit the past? or visit the future?	Would you rather… explore the deep sea? or explore outer space?
Would you rather… have chicken pox? or have stomach flu?	Would you rather… eat ice cream in a cup? or in a cone?

If you could eat one food all week, would you pick... spaghetti? or pizza?	When you eat a cookie with filling in it... do you take it apart? or keep it together?
Do you put ketchup... right on your food? or next to your food?	Do you button your shirt... starting at the top? or starting at the bottom?
Would you rather get... a shot from the doctor? or a bee sting?	Would you rather... be a child? or be a grownup?
When you get dressed, do you put on... both socks, then both shoes? or one sock and shoe, then the other?	Which is harder for you... hopping on one foot? or skipping?

SHAPE RULE CARDS

1

Sort by color

2

Sort by size

3

Sort by shape

4

Sort into 2 groups:

____

NOT ________________

5

Sort by number of sides

6

Sort a new way

Name Date

Student Sheet 7

Calendars

Look closely at a calendar.

Write at least three things you notice about the days and months.

BIRTHDAY GRID

Name _______________________________ Date _______________

Student Sheet 8

Record of Family Ages

As part of our study of data, we are collecting information about ages of the people our families. Fill in the blanks below. For each person, write the name, age, and family role. For example:

My mother Donna is 33 years old.

My brother Andy is 9 years old.

You can decide which people are part of your "family." You don't have to fill in all the blanks!

My ______________________________ is ________ years old.

My ______________________________ is ________ years old.

My ______________________________ is ________ years old.

My ______________________________ is ________ years old.

My ______________________________ is ________ years old.

My ______________________________ is ________ years old.

My ______________________________ is ________ years old.

My ______________________________ is ________ years old.

Name Date

Student Sheet 9

Family Portraits

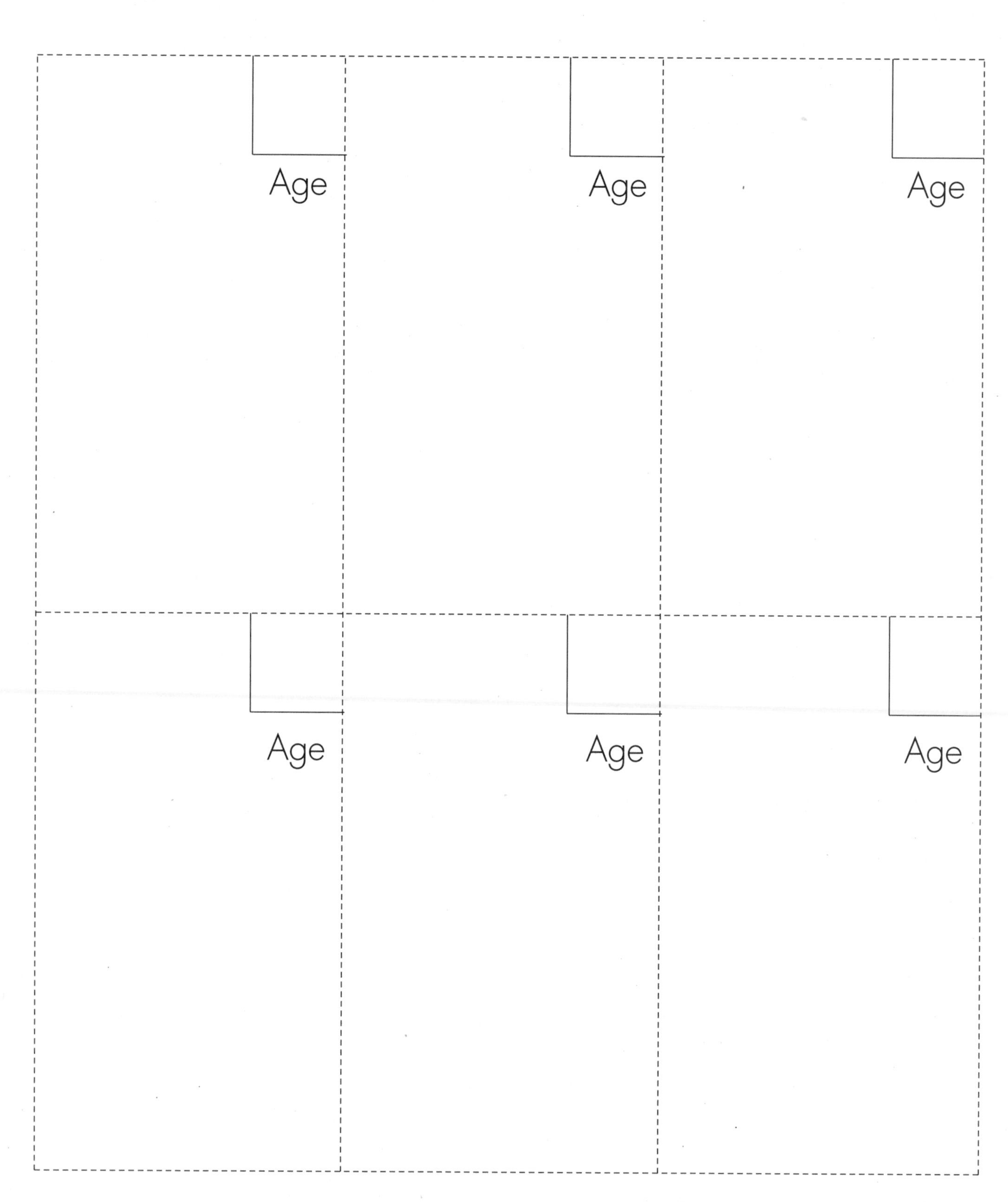

100 CHART

1	2	3	4	5	6	7	8	9	10
11	12	13	14	15	16	17	18	19	20
21	22	23	24	25	26	27	28	29	30
31	32	33	34	35	36	37	38	39	40
41	42	43	44	45	46	47	48	49	50
51	52	53	54	55	56	57	58	59	60
61	62	63	64	65	66	67	68	69	70
71	72	73	74	75	76	77	78	79	80
81	82	83	84	85	86	87	88	89	90
91	92	93	94	95	96	97	98	99	100

Practice Pages

This section provides optional homework for teachers who want or need to give more homework than is suggested to accompany the activities in this unit. With the games and problems included here, students get additional practice in learning about number relationships and solving number problems. Whether or not the *Investigations* unit you are presenting in class focuses on number skills, continued work at home on developing number sense will benefit students. In this unit, optional practice pages include the following:

On and Off; Counters in a Cup These two games are introduced in the grade 1 unit *Building Number Sense.* If your students are familiar with them, simply send home the directions and the game grid recording sheet. Otherwise, plan to introduce each game in class and help students play once or twice before asking them to play at home. If students are very familiar with the game, you might suggest they do the optional step. Ask students to return their recording sheet to class.

Number of the Day For this activity, choose as the "number of the day" a number you want students to practice with, such as 10 or 12, or perhaps today's date. Students write expressions that are equal to that number. For example, if the number of the day is 12, possible expressions include these:

$10 + 2$ $3 + 7 + 2$ $4 + 4 + 4$ $13 - 1$

Introduce the activity in class to be sure that students understand the range of possible responses. Practice Page A can be used more than once; fill in a number before copying it each time you want students to work on Number of the Day at home.

Story Problems Students work with story problems in the grade 1 units *Building Number Sense* and *Number Games and Story Problems.* The problems included here for practice specifically continue the type of work students did in those units, recording their solution strategies using pictures, numbers, and words. If you haven't done either of these units in class, it would be better to assign other practice pages for homework. For additional practice, you can make up other problems in this format, using numbers and contexts that are appropriate for your students.

Name Date

On and Off

> **Note to Families**
> For counters, you might use buttons, pennies, paper clips, or toothpicks. If you do not have a copy of the On and Off game grid, write the numbers in two columns on any paper.

Materials: Counters (8–12)
On and Off game grid
Sheet of paper

Players: 1–3

Object: Toss counters over a sheet of paper. Record how many land on and off the paper.

How to Play

1. Decide how many counters you will toss each time. Write this total number on the game grid.
2. Lay the sheet of paper on a flat surface.
3. Hold the counters in one hand and toss them over the paper.
4. On the game grid, write how many landed on the paper and off the paper.
5. Repeat steps 3 and 4 until you have filled the game grid. (Take eight tosses.)

Optional

Your filled game grid shows different ways to break the total number into two parts. Can you find a way that is not shown?

Name ____________________ Date ____________________

On and Off Game Grid

Total number: _______

On	Off

Total number: _______

On	Off

Name Date

Counters in a Cup

Note to Families

For counters, you can use buttons, pennies, paper clips, beans, or toothpicks. Hide them under any container that you cannot see through. If you do not have a copy of the game grid, write the numbers in two columns on any paper.

Materials: Counters (5–10)
Counters in a Cup game grid
Paper cup

Players: 2

Object: Figure out how many of a set of counters are hidden.

How to Play

1. Decide how many counters to use each time. Write this total number on the game grid.
2. Player A hides a secret number of counters under the cup and leaves the rest out.
3. Player B figures out how many are hidden and says the number. Lift the cup to check.
4. On the game grid, write the number hidden in the cup and the number left out.
5. Players switch roles. Hide a different number of counters. (It's OK to hide the same number of counters more than once in a game.)
6. Repeat steps 2–5 until you have filled the game grid. (Hide the counters eight times.)

Optional

Your filled game grid shows different ways to break the total number into two parts. Can you find a way that is not shown?

Name ______________________ Date ______________________

Counters in a Cup Game Grid

Total number: ________

In	Out

Name Date

Practice Page A

The number of the day is ______.

Write equations that show ways to make the number of the day.

Name Date

Practice Page B

Nora has 6 library books. She took out 8 more library books. Now how many library books does she have?

Show how you solved this problem. Use pictures, numbers, or words.

Name Date

Practice Page C

There were 14 children on the school bus. Then 7 children got off the bus. Now how many children are on the bus?

Show how you solved this problem.
Use pictures, numbers, or words.

Name Date

Practice Page D

Joey had 12 pennies. He spent 10 of them.
Now how many does he have?

Show how you solved this problem.
Use pictures, numbers, or words.

Name Date

Practice Page E

Kim has 4 purple markers. Gina has 5 green markers. Peter has 6 red markers.
How many markers do they have in all?

Show how you solved this problem.
Use pictures, numbers, or words.